Foreword

For centuries people interacted with horses as part of their daily lives. They used them in war, as a means of travel from town to town, as equipment on the farm, as transport to school, and for many other purposes. In the 21st century, the horse is a recreational and a competitive tool and a pet, but not an integral part of our society. This has resulted in a different understanding of the horse's psyche, and widespread ignorance and misunderstanding of the horse's natural behaviour. Anthropomorphism (attributing human characteristics to horses) has crept in, often causing inappropriate human responses to a horse's actions and widespread beliefs such as 'the horse loves me' and 'wouldn't hurt me' because it is a 'good' horse. *Horse Safe* demystifies many of the current ideas of horse behaviour and motivations, and helps horse people see the patterns of natural stimuli that are the driving forces for horses' actions and responses. It covers the significant areas linked to interactions of horses and people in everyday life.

Jane Myers is well-qualified to speak in this field. She has a wealth of experience and knowledge in horse behaviour and responses, and is eminently qualified to write this book on sound philosophies of safe practices around horses. She has a strong background in academic pursuits, has written a number of books on horse care and keeping, has personal experience with horses covering many years, and recently has been involved with the Association for Horsemanship Safety and Education (AHSE) in Australia as a clinic instructor. That role involves her in assessing candidates as instructors and/or trail guides working in areas like trail riding, riding schools and camps, as well as private instructional situations. These workshops focus on the principles behind safe practices in and around horses.

Horse Safe includes the combined wisdom of the many staff who have run AHSE clinics for 15 years, and draws conclusions from joint discussions with over 1000 knowledgeable and experienced horse people.

One of the most outstanding features of AHSE clinics is the consensus of participants and staff that the general public knows very little about keeping safe around horses, and that most horse people either have not thought through the principles of safe horse handling, or knowingly break the existing rules! Horse people often excuse their inappropriate behaviour by one of three rationales. They believe that they 'know what they are doing' and can cut corners and still avoid the dangers, they do not believe their horses would do anything to hurt them, or they are confident that their responses will save them if something goes wrong. There are too many tales of experienced horse people getting badly hurt, even killed, using these rationales for inappropriate behaviour around horses.

Horse Safe will be useful for all horse people: from a child who has just bought their first horse (or is horse-mad and just dreaming of that event), strappers in sporting events, pony club members, jockeys and trainers, casual horse riders, through to serious competitors. It is a significant tool that will add to horse people's understanding of what makes their horse tick, the reasons behind many of the safe handling rules already in place, and the principles that underpin the decisions we should be making to keep ourselves, our loved ones and the public safe as they interact with horses.

AHSE is pleased to endorse *Horse Safe* as a valuable tool for all people who are interested in or who work with horses. It is a significant source of interesting insights into the horse's psyche as well as confirmation of conventional wisdom about the horse, along with myth-busting some common practices and beliefs. A fascinating and enthralling read for all horse people!

Paul Davenport and Nina Arnott
Clinic Instructors, Association for Horsemanship Safety and Education in Australia Inc.

August 2005

Contents

Preface

Horse activities are enjoyed widely across Australia. They make a major economic and cultural contribution to our country as well as being integral to our heritage. Upwards of 250 000 people ride horses each year in Australia, and while the rate of accidents is low in comparison to the total number of participants, serious and preventable accidents do occur. Any effort to reduce accidents associated with horses is to be commended.

To facilitate safe interaction between horses and people, in 2003 with assistance from the Federal Government, the Australian Horse Industry Council (AHIC), a representative organisation serving the Australian horse industry, produced a Code of Practice for horse activities. Commonly referred to as the 'Horse*Safe*' Code of Practice, the document includes requirements and guidelines for managing horse activities. Details of the Code and the AHIC can be found in Appendix 2 and at www.horsecouncil.org.au.

The publication of this book on horse safety is applauded by the AHIC. Indeed, to assist with the spreading the message of safety around horses the AHIC agreed to the use of the Horse Safe words for the title. However, readers should not confuse this book with the Horse*Safe* Code of Practice. This book is a 'how to' guide aimed largely at educating individuals on how to behave around horses. Conversely, the Code of Practice, while being useful for individuals is largely aimed at the businesses and organisations that run horse activities.

The AHIC is confident that practicing the principles in this book will make the reader a better and safer horseman or woman. After all, safe horsemanship equals good horsemanship.

Paul O'Callaghan
Veterinary Surgeon
Racing Victoria Limited
President of the Australian Horse Industry Council (AHIC)

September 2005

Acknowledgements

Thank you to the following people for reading and commenting on the manuscript:

- Nina Arnott, AHSE
- Marguerite Bongiovanni, Victoria
- Paul Davenport, AHSE
- Jim Doig, South Australia
- Sandy Doig, AHSE
- Liz Foster, AHSE
- Fiona Glenn, Victoria
- Norm Glenn, P.C.A.V. State Councillor
- Russell Mathews, AHSE
- Jamie McLean, AHSE
- Margaret Mooney, AHSE
- Jane Williams, Equine Studies lecturer, Glenormiston Campus, University of Melbourne
- Noel Wiltshire, AHSE

Thank you to the following people for providing photographs, modeling or supplying equipment:

- AHSE
- Steve, Amy and Chad Brady
- Gold Coast Polo and Country Club
- Kristen Downs
- JR Easytraveller Floats
- Annie Minton
- Aarron Myers
- Lauren Myers
- Stuart Myers
- RDA Crowson Park, Queensland

Introduction

Horse riding and its related disciplines are widely acknowledged as being high-risk activities. Due to the high accident figures, insurance companies and the like continue to pigeonhole horse activities as extreme sports. This is not surprising when the Australian statistics are taken into account. Between 2001 and 2003 there were an estimated 2400 hospital admissions due to horse-related injuries. In the same period approximately 30 people died as a result of their injuries (R. Cripps, pers. comm. 2005). Equestrian activities have among the highest risks of serious injury and death of any sport, including motorcycle and car racing (Sport & Recreation Vic.). About 80% of horse-related injuries occur when riding (falls) and the rest occur when handling and around horses (kicks, crushing, bites etc.) (Sport & Recreation Vic.).

It will never be totally safe to interact with horses because horses are large animals with their own instincts and agenda and we tend to put them into unnatural circumstances. However, many of the dangers can be reduced and managed by employing safe practices and risk management strategies. These factors need to be incorporated into all dealings with horses, by all levels of horse people from complete beginners to advanced horse people. Horse people need to be responsible for their own safety and for making sure that they do not put other people at risk. Horse people need to be aware that non-horse people are usually unaware of the potential for injury when around horses, and take steps to keep them safe.

Many horse-related injuries and fatalities are due to lack of understanding about why horses do the things they do, lack of experience, and because we get complacent. What we consider to be irrational behaviour is not irrational to a horse. Being safer around horses may mean learning new skills or changing your ways of thinking and behaving, using safe practices and risk management strategies. These practices and strategies require horse people to:

- learn about the physical and behavioural characteristics of horses in order to improve understanding and use this knowledge to be safer when around horses;

- use safe equipment and facilities, improve existing facilities when necessary and be proactive against the use of unsafe equipment and facilities;
- manage horses correctly so that they can perform well, willingly and without the resistance that can arise from poor management;
- use safe procedures when handling, riding and training horses and train horses to be safer to handle and ride;
- have and provide good safe instruction. Self-improvement and the improvement of others reduces accidents;
- be vigilant and aware of the environment and surroundings. This involves identifying hazards, acting on them if necessary and using peripheral vision to notice any changes that may cause an accident;
- become safety-conscious to the point where this becomes second nature;
- expect the unexpected when dealing with horses and be prepared but not nervous;
- never do anything that they are not comfortable with. Many accidents are caused because people, through either inexperience or bravado, attempt to do things with horses that are way above their or the horse's level of skill;
- be aware of other people (especially children) and be conscious of the possible reactions of their horses that could put others at risk!
- be prepared to take control if a situation becomes potentially dangerous, e.g. if a handler/rider is concerned that the activity may involve a risk that they cannot control, they should stop what they are doing and ask people to move etc.

This book gives an objective view of safety issues with horses, whatever your chosen discipline or current skill level. While recognising that there are numerous opinions in the horse industry about how to do things with horses and many different methods of handling and training horses, the safety issues remain the same. Therefore fundamental safe practices and risk management strategies can be applied to any situation with horses. Whether you are new to horses or have been around them for some time, whether your involvement is for personal reasons or commercial, the information in this book will help to improve safety, reduce risks and prevent accidents for people that handle and ride horses and for people that are not directly involved but could still be affected by them, such as bystanders and spectators.

Each chapter deals with its subject from a safety standpoint. By learning about the fundamental safety aspects of various situations, we can learn how to make informed decisions about safety in our own circumstances. *Horse Safe* is not intended to be a horse care/management manual and consciously omits such references. Each chapter includes recommended reading that will broaden your knowledge of that particular subject, if required.

The information in this book is aimed at all horse people from experienced to complete novices and from those responsible for their own safety to those that are responsible for the safety of others, such as instructors and parents. It is written in plain language so that everyone can understand it, but some of the information is relevant to more experienced people and certain procedures should only be attempted by them. When this occurs, I highlight the necessary level of experience.

It could be argued that if inexperienced horse people don't 'have a go' at doing things with horses that are above their current skill level they will never learn new skills. However, if you are an inexperienced horse person you must learn the basics before you can progress to procedures that involve more skill. People must learn to walk before they can run! The biggest danger with inexperience is that you are not aware of how much you do not know. Many experienced horse people look back with wonder at some of the things that they got away with (and some that they didn't) as they were learning, and would not advise novices to learn the same way that they did.

As stated previously, there is no totally safe way to interact with horses. However, by raising our awareness of safety issues and becoming proactive in dealing with them we can dramatically reduce the levels of risk.

1

Horse characteristics

This chapter describes the main physical and behavioural characteristics shared by domesticated horses and horses that live in the more natural environment. Given the opportunity or circumstances, domestic horses can and will revert to natural behaviour. In many countries, domestic horses that have escaped or been released have been able to survive and thrive largely because they still possess natural behavioural characteristics. When living in a herd situation, domesticated horses interact with each other the same way that they would in the wild. The main difference is that wild horses that have not had contact with humans or that have had only bad experiences with humans act instinctively. Most domesticated horses have been habituated to accept humans and various human-related sights and sounds and have been trained, to a lesser or greater degree, to override their instincts when being handled and ridden. Domesticated horses can and often do, however, revert to instinctive behaviour in some circumstances depending on their individual behavioural characteristics and their level of training. Even a well-behaved and well-trained horse can react instinctively from time to time when the stimulus is strong enough. Therefore, it is important that people who deal with horses understand horse characteristics.

This chapter describes the horse characteristics that may affect your safety when associating with them. Later chapters explain how to make interactions with horses safer, by taking these characteristics into account.

Physical characteristics

Horses are large, grazing, herd-living herbivores. Domestic horses have been bred to range in size from less than a metre to 1.8 metres at the withers. A large horse can weigh 1000 kg; even an average size (15hh) horse usually weighs around 500 kg. A foal weighs around 50 kg. Apart from size, all horses share the same characteristics.

A horse has very sensitive skin, which is necessary for detecting flying pests and parasites. It can twitch the skin in many areas of its body and will even kick forwards with a back leg if it feels the light touch of an insect or something it perceives to be an insect.

Some parts of the horse are even more sensitive than others. These include the top lip, muzzle, ears, feet, legs and flanks. The top lip has many nerve endings and is highly mobile, allowing the horse to investigate, feel and select with it. The muzzle is covered in whiskers that help the horse to 'feel' in much the same way as a cat.

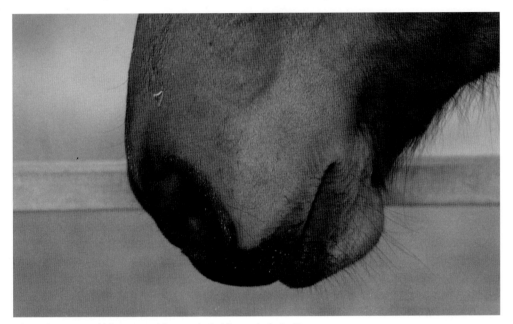

A horse has many whiskers around its muzzle that it uses to feel with

The eyes, ears, feet and legs of a horse are essential to its survival and a horse will endeavour to protect there areas when under threat. In some mares the flanks are sensitive to the extent of being ticklish, as the flanks are sniffed by a stallion prior to mating to test the mare's receptiveness. A horse has a strong sense of smell. It uses this when assessing a potentially dangerous situation.

The teeth are sharp and the jaws are extremely strong. In addition, a horse's long neck makes it easy to swing the head from side to side or reach out to bite. A bite from a horse causes a crushing rather than a tearing injury (as is the case with a dog bite). Horses bite or nip each other frequently when playing or asserting themselves.

A horse's hoof is relatively small and light, compared to a human foot, on the end of leg that is concertinaed to allow a lot of extension when running and kicking. This means that very little energy is required to move it and it can move very quickly. It also means that a horse can kick with accuracy and strength. A horse can move this relatively lightweight

Horses bite when playing or asserting themselves

hoof quickly and easily and, because there is a small area of contact, like stiletto heels on a soft floor it can cause a lot of damage.

A horse is capable of striking forwards with a front leg. In the wild, a horse that is surrounded by predators will strike at them with a front leg or rear to use both front legs to strike and stamp if necessary.

Horses can kick out with one back leg …   or with both back legs

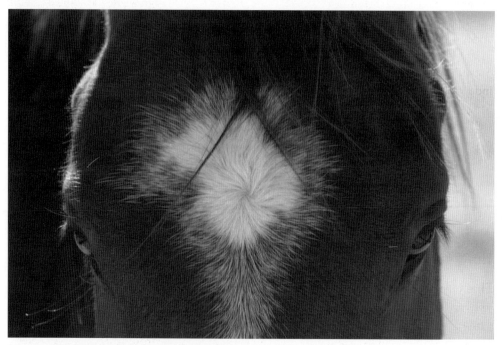

A horse's eyes are on the side of the head, which makes it difficult to see straight forward

A horse's eyes are situated on the sides of its head so that it has very good all-round vision at the expense of sharp forward-focusing vision. The horse can detect moving objects in the distance that we would struggle to see. A horse also notices when there are unfamiliar objects in a familiar place, because horses take the whole situation into account when looking for potential dangers. A horse can't focus on objects as quickly as we can, and this lack of sharp focusing can cause it to be anxious about an object until it has time to see the object better and identify it. If the horse can't very quickly ascertain what the object is, it prepares for flight.

When its head is in the grazing position the horse can see along either side of its legs and, with small movements of the head, can see directly behind and in front. If the horse identifies a possible danger it immediately raises its head above its body, where it can again see all round by moving its head slightly. The only time a horse is unable to see behind it is when the head is level with the body. At all times, a horse has a blind spot directly in front of its face and it can't see the area below and behind its chin.

A horse can listen forwards and backwards at the same time

The horse's ears are on top of the head and are serviced by numerous muscles. This means that each ear can swivel 180°, giving potential for all-around 360° hearing without having to move the head. The ears work independently – a horse can have one ear facing forwards and one backwards at the same time. This lets it hear very well indeed – a horse can hear things approaching from behind well before we can. When riding, you can tell, what direction the horse is looking by looking at the ears. The ears also form part of the horse's body language.

Many horses are unsettled in windy weather. One of the reasons for this is because the wind distorts and adds to the normal noise level, making it more difficult for the horse to detect potential danger. In addition, horses do not like wind and rain together and will turn their rump into the wind to protect the head.

Behavioural characteristics

Flight response

Horses utilise the 'flight response' if they are frightened, because they are prey animals rather than predators. Horses are highly reactive compared to other prey animals, some of which are adapted to defend first and then, if that fails, flee from danger. Horses resort to defence only if they are captured or cornered, such as being caught by predators or in a confined area. The physical and behavioural characteristics of horses ensure that they act on instinct, to run away from danger or defend themselves if trapped. They will kick, strike or bite if they can't get away from a perceived danger.

A horse is either in an alert state or can become alert very quickly. Even a dozing horse can switch to being alert in an instant if an unfamiliar noise or movement occurs or if other horses nearby become alert. This is part of the natural behaviour of an animal that is hunted by other animals.

Horses run away from danger as their primary reaction to being startled

When you are riding or handling a horse you must have its attention, but when a horse is in a highly alert state it doesn't listen to its rider or handler. A horse is most dangerous when it is frightened because it won't respond to cues as it will when calm. There is a real danger that the horse may panic, spin away from the perceived danger, and run. Good training results in the handler or rider being able to take control of the horse before it becomes highly alert. Good training also 'habituates' the horse to many situations so that it does not become highly alert (see Chapter 8) as easily.

Herd behaviour

A fact about horses that cannot be ignored is that they are herd animals with a strong herd instinct that drives their behaviour, because it is highly instinctive for horses to want to be together. Living in a herd means safety in numbers: it means that each animal reduces its chances of being caught by a predator and there are many eyes and ears looking for predators. As horses spend a lot of time grazing with their head down in the grass, which reduces visibility, this point is very important. A horse on its own in the wild is much more likely to be caught by a predator and it expends too much nervous energy having to remain in an alert state. Horses that live in herds can take it in turns to be alert, to eat and to rest.

When allowed to live as a herd, domestic horses, like their more natural-living cousins, have a rich social life which includes such behaviours as mutual grooming sessions, playing, grazing and simply standing around together.

The fact that horses are herd animals rules their behaviour and can cause them to do things that seem irrational to us, such as panic if they get separated from other horses.

Social hierarchy and aggression

The social hierarchy within a herd is maintained with a whole array of body language and interactive behaviours, for example small gestures such as ears back and head thrusts which usually result in the subordinate horse moving aside. This behaviour maintains order in the group and is why herd or pack animals are much easier to train than animals that are solitary by nature. Herd animals are used to and feel secure in a structured environment, where their role within the group is directed by a social hierarchy.

A horse that is confident in its dominance can move other horses using as little effort as possible. If you watch horses in group situations you will see that small gestures can get big results. Generally, only horses whose status is questionable use overtly aggressive gestures.

A dominant horse drives another horse away by laying the ears back and 'driving' with the head

In the natural environment, horses are not particularly aggressive animals. Their food source is widespread and they do not have to fight each other for it. Aggression is used only when absolutely necessary because it is dangerous for the aggressor as well as for the victim. For example, a stallion may become aggressive if he is challenged by another stallion, but the remainder of the time he is often subordinate to the lead mare. In a domestic situation horses are mainly aggressive towards each other at feed times. They can injure people and each other if they are not separated when feed is being handed out, as a concentrated feed source creates competition.

Intelligence and learning ability

It is important not to give horses human characteristics (anthropomorphise), but to remember to view their behaviour from their perspective. What seems logical to them may seem illogical to us. Beware of attaching human emotions to horses as this may impede understanding of their behaviour. For example, horses do not hold grudges, unlike most humans! They accept or may challenge their position in the group but they don't sulk if they can't have their own way.

The problem is, when we judge intelligence in animals we usually measure it from a human perspective. Humans can be logical, good problem solvers and able to reason and think things through. This is because our brain is very large in relation to body size, and the part of the brain that deals with tasks such as reasoning and problem-solving is particularly well-developed. A horse's brain is different and they can't reason or solve problems other than by trial-and-error. So, trial-and-error learning is a process commonly and successfully employed by good trainers.

An example of trial-and-error learning is when a horse plays with a gate-catch, one day it happens to open and the horse is rewarded by getting to the grass outside the gate. If the horse does this a few times it learns that playing with the gate-catch results in a reward. The horse didn't originally set out to open the gate, but that was the result and so it seems that the horse used problem-solving skills although in fact it didn't.

This doesn't mean that horses are not intelligent; they simply have a different kind of intelligence from humans. Therefore they often do things that seem remarkably stupid to us. For example, if a rider falls off a horse but their foot is caught in the stirrup, the horse doesn't reason that the rider can't get free and so if it stands still the rider will be able to untangle their foot and all will be well. Instead, the horse can panic at the unfamiliar sight and feel. To the horse, the rider has become a potential threat and unless the horse has been trained otherwise it will panic and bolt.

Even though horses are highly reactive and nervous they can relax when in familiar surroundings and can learn to accept familiar sights and sounds. This ability to relax in familiar surroundings is a necessary behaviour for a prey animal, otherwise they use too much energy remaining alert all the time. Horses can also generalise sights and sounds to some extent, therefore the more they see and do the more they accept as 'normal'.

Horses have a very high learning ability and can learn certain tasks very quickly if trained properly. Not only does a horse learn quickly, but it will remember and respond to a cue indefinitely once it has been taught thoroughly. This of course includes negative as well as positive responses.

A well-trained and relaxed horse will move away from any pressure applied to its body, for example if you put your hand on its hip it will move away from your touch. An untrained horse will either ignore the pressure or may even lean back into the pressure. This can result in a person getting crushed against a solid object. It is through training that horses learn to yield to pressure cues, one of the most important aspects of training a horse (see p. 125).

Body language

Horses use body language to convey their intentions to other horses. Even though horses are highly reactive they often give lots of signs before acting. Different parts of the horse's body are used to communicate different signals.

The ears are used in many ways to signal a horse's intentions or its state of mind. When a horse is very relaxed it may allow its ears to 'flop' sideways. Horses vary in how much they can do this. They will do it when standing dozing but may also do it when being ridden, especially if they have become soft and relaxed in the jaw.

Make sure that a dozing horse is aware that you are there before you approach it, as otherwise it may be startled. Wait until the horse has turned an ear or its head towards you before you approach.

Horses 'prick' the ears forwards and lift the head when intent on listening forwards

Horses lay the ears back against the neck when they are threatening

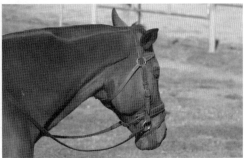

Compare this photo to the one above right. Here the horse is listening behind but is not laying the ears back aggressively

The horse is dozing and the ears are back. Again, it is not an aggressive gesture

Generally, when a horse is alert the ears point towards whatever the horse is looking at. A horse will prick both ears forwards and lift the head when it is very alert. Pricked ears mean that the ears are as far forward as they can go, as the horse is intently listening forwards. Very pricked ears can mean that the horse is no longer listening to the handler or rider.

Both ears pressed flat back against the neck is a threat. Flat back ears are usually accompanied by wrinkling of the skin above the nostrils and sometimes below the eyes. This threat is commonly used between horses, however, it can also be used as a threat to humans.

Both ears pointing back but not pressed to the neck mean that the horse is listening behind or that the horse is resting and has allowed the ears to relax. You can tell the difference by looking at the horse's facial expression. If the horse is listening behind it will look alert and have the head higher. If it is dozing it will have half-closed eyes and a lowered head.

The tail can give many signals. It is used to swish away flies but it is also swished in irritation. A horse clamps its tail to its rump when frightened and sometimes when about to kick or while actually kicking. Sometimes a horse will also tuck its rump under when it clamps its tail.

A relaxed horse will carry the tail slightly lifted when moving (Arabians naturally carry it higher than other breeds), without swishing. If the horse is standing still the tail will be touching the body but relaxed, i.e. not clamped. An excited, exuberant horse will sometimes carry the tail up and over the back or straight up in the air like a flag.

A horse carries its tail like a flag when it is excited

The mouth area gives off many signs. The chin and lips become bunched and tight when the horse is tense. The lower lip droops when the horse is relaxed. A young horse 'claps' its lips to older horses to signify its subservience. A horse curls its top lip when it smells something strange (Flehmen response). Stallions and some geldings also do this when sniffing a mare in oestrus. A horse flares its nostrils and snorts when alert.

A horse uses the head in a forward-thrusting motion to drive another horse out of the way. Some horses, particularly stallions, swing the head around in a complete circle when they are feeling exuberant. This is usually done when the horse is running around.

When a horse is planning to kick it will often raise a back foot off the ground and hold it there as a threat. This will be accompanied by other body language signals such as ears back. The horse may also swing the back end towards the object it is planning to kick.

A horse that is 'resting a leg' is relaxed and may be dozing. However, the resting-leg stance can mean that the horse may pick up that leg to kick more quickly.

The Flehmen response is shown when a horse smells something unfamiliar

A horse lifts a foot off the ground as it plans to kick

The body can be tight and tense or soft and relaxed depending on how anxious the horse is. Horses naturally lean into pressure rather than move away from it until they have been trained otherwise. When two horses kick each other they press hard against each other or spring right away. This is because a kick is most dangerous when the leg is fully extended, therefore the closer the horses are the less impact the strike will have.

When a horse is highly alert or frightened it usually freezes and lifts the head as high as possible. If eating, it stops chewing so it can hear more effectively. The eyes and ears are fixed on the object in question (see p. 7).

All these body language signals can be used by observant humans to decipher what the horse is likely to do next. It is important to have a good understanding of the horses' body language to take advantage of this early-warning system. This gives you time to respond rather

A horse can rest one back leg while dozing

than having to react to situations. Once you are experienced at reading horses' body language, you can act by either moving quickly to a safe spot or by handling the horse in a way that changes the behaviour.

These horses are highly alert

Vocalisations

Horses make several different types of vocalisation.

- Loud snort: horses snort hard and loud when they are highly excited or when they are looking at something that they cannot identify.
- Soft snort: horses snort more softly when less aroused. Some horses make a snorting noise when cantering or galloping, which is actually the vibration of the inner nostril.
- Squeal: horses squeal when greeting one another and when on the defence or as a sign of aggression, e.g. when kicking.
- Whinny: horses make several noises that can be classed as a whinny, ranging from high-pitched and urgent to a more relaxed greeting whinny.
- Nicker: horses make a low rumbling noise that is used as a greeting.
- Nose blowing: horses blow their nose when relaxed.
- Sigh: horses sigh when relaxed.

Horses do not squeal or make any other vocalisation to signal pain, unlike a dog, for example, that will yelp and whimper.

Gender differences

There are some differences in behaviour between male and female horses but the difference is not as extreme as many people believe, especially between geldings and mares. In a natural living situation when a herd consists of a stallion, several mares and their offspring a mare is usually the leader of the herd. The stallion fights off other stallions or colts that attempt to claim the harem, but is otherwise usually subservient to the lead mare.

Geldings are castrated domesticated male horses. People generally believe that geldings are the most tractable of the three groups of horses but, as with most rules, there are exceptions. Some geldings exhibit some sexual behaviour, especially during spring when hormones levels are rising. Most geldings show no interest in mares but a few show so much interest that they mount mares and act aggressively with other geldings in the area. This behaviour can be a result of being incorrectly or incompletely gelded (classed as a 'rig' or more correctly a cryptorchid). Rarely, the horse was correctly gelded but can still secrete the hormones that cause him to behave like a stallion (without being fertile). A vet can determine the situation.

The behaviour of mares varies. A small number of mares show a distinctive change in behaviour when they come on heat. These mares are not suitable for inexperienced handlers or riders. Many mares show very few, if any, behavioural changes. Signs that a mare is on heat include urinating, lifting the tail, squatting the hind quarters and standing with the hind legs spread when stallions or geldings are around.

A mare with a foal is usually very protective of the foal and can become aggressive if she feels she has to defend the foal. Mares and foals become very agitated if they are separated, to the extent that they can run through or jump over anything in the way in order to get back together.

Domesticated stallions range in behaviour from being as tractable as a well-behaved mare or gelding, to being difficult to manage. Unruly behaviour is usually due to a lack of understanding and poor handling and training.

Food and eating behaviour

Food is very important to horses because in the natural situation foraging takes up most of their time. Horses eat for up to 20 hours a day if necessary, such as when eating low-energy grass. They will eat for around 12 hours per day on good-quality grass. Chewing is also important for the production of saliva that buffers acid in the stomach. When a horse is chewing it is in a more relaxed state, especially if it is eating with the head down, such as when grazing.

In the natural situation chewing is done with the head in a lowered position (as opposed to eating out of a raised feeder, for example), so the horse connects the head-down position with being relaxed. This is one of the factors that we can use to help train horses (see p. 125).

Grazing is very important to horses

Individual behavioural characteristics

Even though all horses share the same basic behavioural characteristics they also have individual differences. Some are more reactive than others, in some situations. Individual differences can be due to one factor or a combination of several, such as age, natural differences in temperament and different training and life experiences. The individual characteristics of a horse must be taken into consideration for safer handling or riding of that horse.

Recommended further reading

Budiansky, S (1997). *The nature of horses*. Weidenfeld & Nicolson, London.
Kiley-Worthington M (1987). *The behaviour of horses*. J A Allen & Company, London.
McGreevy P (2004). *Equine behavior*. Saunders, Sydney.
McLean A (2003). *The truth about horses*. Penguin Books, Melbourne.
Mills D and Nankervis K (1999). *Equine behaviour: principles and practice*. Blackwell Science, London
Rees L (1984). *The horse's mind*. Stanley Paul, London.

2

Human physical requirements

Anyone who handles or rides horses without assistance should have reasonable to good fitness, flexibility and strength depending on their level of involvement. Fatigue can reduce reaction time, resulting in you being unable to move out of the way when necessary, or reduced concentration. It also leads to frustration, which is a dangerous emotion where horses are concerned.

People who are involved in caring for horses need to be fit and strong enough to carry heavy objects such as water buckets, feed, hay and bedding, and to be able to walk reasonable distances at the same speed as a horse. At the same time, caring for and riding horses can raise your fitness level, resulting in better health.

Other forms of exercise such as walking, cycling and swimming, in conjunction with horse care and riding, can further increase coordination, fitness, flexibility and strength and result in fewer injuries. Exercises such as Pilates and yoga help to build the core muscles that are essential for riding. Stretching before riding warms up your muscles and reduces the chances of injury. Gymnastics and martial arts such as kung fu, judo, ju jitsu and taekwondo have the added benefits of teaching you how to fall better as well as improving your coordination, fitness, flexibility and strength.

People with mobility problems, back problems and/or injuries that make strenuous work difficult should be assisted by more able-bodied people when caring for and riding horses. A doctor should be consulted before commencement. Sports physiotherapists or sports masseurs can help many problems, so it is always a good idea to consult a qualified person to see if your condition can be improved.

Everyone that cares for horses should practise safe lifting techniques when lifting heavy loads such as feed sacks and when lifting a ramp on a trailer. You must keep your back straight and bend your knees when lifting. Heavy loads should be shared or you should use some form of trolley.

People who come into contact with horses on a regular basis are usually advised to be regularly inoculated against tetanus (consult your doctor).

Riding while pregnant is a personal choice. It is advisable to check with your doctor before doing so; their recommendation will probably take into account your level of skill. The Sex Discrimination Act 1984 means that in most circumstances women cannot be banned from taking part in sports while pregnant.

Recommended further reading

The national EFA website (www.efanational.com) has risk management information including a discussion of riding while pregnant.

Dennis DR, McCully JJ & Juris PM (2004). *Rider's fitness program: 74 exercises and 18 workouts specifically designed for the equestrian.* Storey Publishing, US.

Targett T (2005). 'Baby on board'. *Hoofbeats Magazine*, Volume 26, no. 5, Feb/March.

3

Clothing and equipment

This chapter describes what to wear and what not to wear from a safety point of view when handling and riding horses. Clothing should also be comfortable, because if it isn't you may become distracted by the discomfort rather than focusing on the job in hand, and therefore operate less effectively.

Helmets

Helmets save lives. This has been proved countless times yet some horse people are still reluctant to wear one. Many people don't like to wear helmets, believing them to be uncomfortable or hot. Modern correctly fitted helmets are neither. Sadly, some high-profile professional horse people don't wear a helmet, giving a bad example to less experienced or younger riders. For example, many high-level dressage riders avoid wearing helmets when training at home. Indeed, at present dressage riders are required to wear a top hat (not a safety helmet) at the higher levels of competition, sometimes leading newcomers to believe that helmets are not necessary. Within western circles helmets are often frowned on and a rider may even be penalised in some competitions for not wearing a Stetson. These cultures are gradually changing as more information becomes available about accidents and safety equipment. For example, in some parts of the world (such as the state of Victoria in Australia) the law requires a horse rider under 18 to wear a helmet when riding on the roads (Vicroads website).

The most common injury for people involved with horses is to the head, both when mounted and unmounted. Head injuries commonly occur when falling from a horse but also from being kicked in the head. They account for the majority of deaths and severe injuries in horse-related accidents (Sport & Recreation Vic.). When you fall off a horse and your head hits the ground and stops suddenly, your brain continues to move, first hitting one side of your skull and possibly bouncing back to hit the other side. A kick to the head can also have devastating results. Even a small bump to the brain can cause concussion; a large impact can cause permanent brain damage or death.

Standards

Helmets should comply with the current standards. In Australia and New Zealand the current safety standard is AS/NZ 3838 (as at 2005). Helmets that meet international standards (such as EN 1384, the current European standard, and ASTM F 1163, the current US standard) are also accepted by some clubs or official bodies. If you are a member of a club or an official body (such as the EFA/pony club, racing board) and you ride under their rules check the rules before buying a helmet.

A correctly fitted, current standard helmet

The standard should be marked on the helmet along with the date of manufacture. Helmets have a shelf life of approximately five years from the date of manufacture, irrespective of when the helmet was purchased. Never buy a second-hand riding helmet – it will have a shorter shelf life and may have damage that you can't see.

Note that bicycle helmets (or helmets from any other sports) are not safe for handling and riding horses. They are designed for impact from a different height and speed, they cover less of your head and they are less stable in a fall than a good equestrian helmet is. A good helmet is not expensive considering how important it is; don't compromise on safety.

Handling horses

There is a large variation in opinion about whether people should wear helmets when handling horses. Many head injuries are caused by horses when a person is on the ground (not riding), such as kicks or strikes to the head or even simply a knock when a horse moves its head quickly. Many commercial establishments insist that children wear helmets at all times on the premises. Many people who handle stallions (when breeding) and young colts (colts have a tendency to rear easily) wear helmets; if they don't they should certainly consider it. Routines such as administering worming pastes, injecting horses and tending wounds are other occasions when you should consider wearing a helmet. Loading and unloading horses into transport is another occasion when handlers are at higher risk of head injuries. Consider wearing a helmet when handling horses as well as when riding. Consider making it a rule/policy that any children in your care/ under your instruction wear a helmet when unmounted as well as when mounted.

Helmet care

Helmets must be looked after carefully to remain safe. Clean a helmet with mild soap and water only. If a helmet is at all damaged or has been knocked against a hard surface, for example dropped on a hard floor or worn in a fall (see p. 177), it should be replaced. Store the helmet where it will not fall on concrete if dropped, and away from direct sunlight. The discarded helmet should be destroyed, not given or sold to another person. It is not possible to test a helmet to check whether it is still safe without actually damaging the helmet. Some manufacturers will replace a damaged helmet for free if it is returned to them, as they can use it for research. Check with the manufacturer.

The only acceptable way of marking the outside of a helmet (as recommended by most manufacturers) is xylene-free texta, not paint or other substances.

Fitting a helmet

A helmet should fit your head snugly and shouldn't slip when you move your head. To fit the helmet, place it on your head to cover your forehead from just above the eyebrows. The helmet should move your scalp when it is rocked backwards and forwards. Long hair should be fastened back but not bunched under the helmet. Never wear anything under the helmet, including hairbands and hairclips. Never put a cap or beanie under a helmet to make it fit better. There should be no gaps around the sides of your head. The harness should be adjusted so that it is snug, not loose.

Recommended further reading

Standards Australia has a website (www.standards.com.au) with information on
    standards for helmets, including riding helmets.
The national EFA website (www.efanational.com) has risk management information
    including helmet requirements and a reproduction (with permission) of
    Appendix C, Helmet Information, from Standards Australia.

Footwear

You will need both workboots and riding boots for working around and riding horses. Boots for handling horses and doing horse chores must have soles with grip, for use in slippery conditions. They can have either inverted or external ridges. These help prevent you slipping and falling under or into a horse when handling or doing horse chores, such as carrying water. Workboots are safer if made from strong leather rather than rubber. Good workboots also help to protect your feet from injury if a horse stands on your toes. Steel-capped boots are safe **only** if they are approved for the weight of a horse. If not, they can buckle under the horse's weight and cause much more damage than the weight of the horse alone. Your toes can be cut off by steel toecaps that are not strong enough to take the weight of a horse.

Sandals or bare feet are not safe for handling horses.

Riding boots can be short or long and are usually made of leather or rubber. They should have a moderate heel of about 2 cm. The heel will reduce the chances of your foot sliding all the way through the stirrup. This can happen while you are riding (and may cause you to lose balance and fall off) or as you are

AHSE

Workboots with external ridging on the sole

AHSE

Smooth-soled, heeled, short riding boots

falling off. If you fall off a horse and your foot is trapped in the stirrup you can be dragged. The results are usually disastrous.

The sole of a riding boot should be smooth, to reduce the chances of the boot getting wedged in the stirrup, or have inverted ridging. Inverted ridging means you still have some grip if you have to dismount.

Boots with external ridging on the sole (such as many workboots) are dangerous in open-fronted stirrups. You can buy stirrups with caged closed fronts, or devices such as Toestoppers™ that turn an ordinary stirrup into one that your foot cannot slide through (see p. 85).

It is not safe to ride in bare feet, pumps or sandals, even when the stirrups have closed fronts. Besides the fact that you have no foot protection when you dismount, your ankle is not sufficiently supported for a secure leg position when riding.

A long rubber riding boot in a caged stirrup

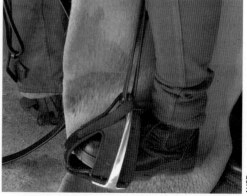

AHSE

A short riding boot that has inverted ridging on the sole

Toestoppers™ can prevent your foot from getting caught in the stirrups if you fall off. They can be fitted to existing stirrups

Short riding boots that are elastic-sided have the advantage that they *may* come off your foot if they get caught in the stirrup during a fall, thus releasing you. Some riding boots have lace-up fronts. There is a danger that the knot or bow can catch in the stirrup iron, particularly with an English type of stirrup iron. Tying laces around the back of the boot makes them safer if they are likely to get caught in the stirrups.

Clothing

Clothing for handling horses should be appropriate for the climate and comfortable to wear. Items also need to be hard-wearing if they are not to wear out too quickly. Considerations for what to wear on your top and bottom half are pretty much the same as for riding, and everyday riding clothes are fine for handling horses. Clothing for riding should be comfortable and not too restricting. Temperature extremes can make you uncomfortable and can result in heat stress or even hypothermia. Good saddlery stores have large selections of riding wear, but camping and outdoor sports shops are also a good source of clothing for handling and riding horses.

Some jeans are fine for riding and it is possible to buy purpose-made jeans. Jeans are not ideal in the rain because they hold water when wet and your legs will stay cold. Any trousers or jeans for riding should not have a bulky seam on the inside of the leg, as this can rub. Flared jeans are unsafe because they can get caught in the stirrup leathers and can cause tripping on the ground. Jeans that are too tight cause discomfort and lowcut hipster jeans mean that you cannot concentrate properly because you have to keep hitching them up.

Pantyhose under jeans or pants can prevent rubbing, particularly on men with hairy legs!

Jodhpurs are designed for riding and if well-fitted are very comfortable. Shiny trackpants are too slippery and tend to wear out too quickly. Basically, pants or leggings that are made from heavyweight stretch fabric, are close-fitting, have no or minimal seams on the inside leg and are long enough to cover your entire leg (without 'riding up') should be comfortable enough for riding.

Rider wearing a helmet, jeans, smooth-soled boots and a shirt

Rider wearing a helmet, jodhpurs, short smooth-soled boots, short leather chaps and a waistcoat

Leather, suede or waxed cotton chaps can reduce rubbing on the inside of your calf when riding. Long chaps were originally designed to protect the legs from sharp thorns etc. when riding in rough country. Short chaps finish just below the knee and protect only the lower leg. Long waxed cotton chaps are useful in wet climates for rain protection. They also add extra warmth in cold weather.

Short leather chaps

A helmet fitted with a sun brim

Rider wearing helmet, short smooth-soled boots and long chaps over stretch pants

You should wear at least a t-shirt or shirt with sleeves when handling or riding a horse. If you're going out in the sun, a shirt or top with long sleeves and a collar that can be turned up is necessary for sun protection. Always make sure that any exposed skin is protected by sunblock. You can buy a slip-on wide brim that fits over your helmet for sun protection, which is a useful addition in hot weather.

Waistcoats or vests are useful because they keep your body warm while allowing free arm movements. Waistcoats allow you to maintain a more comfortable temperature without having to remove clothing so often, which is helpful when riding. Never put on or take off gear while mounted.

In cold or wet weather, clothing should be warm and waterproof. Heavy oilskin coats tend to be too restrictive and are only necessary in very heavy rain and on longer rides. A lighter-weight outdoor waterproof jacket is more comfortable for shorter riding sessions as they tend to be warm as well as dry. Separate waterproof trousers, waterproofed suede leather chaps or waxed cotton chaps can also be worn to keep your legs dry. Make sure that your horse is habituated (see p. 128) to the sight and sound of any new materials, such as rainwear before riding in such clothes.

Gloves can be worn when riding and handling horses. Some horse handlers and riders find it difficult to use gloves because they can reduce 'feel'. Gloves are invaluable when riding in wet weather, to maintain grip on wet leather reins. They must fit well – if they are too big you may have problems managing reins or leadropes.

Riding clothing for competing varies from sport to sport. Check what is required with the club or official body.

Protection

Body protection

A body protector

Body protectors are becoming more popular with riders. At present they are compulsory in Australia for jockeys and track riders. In other horse sports they are often recommended but are not always compulsory: check the rules of your horse sport and consider wearing one even if not compulsory. Their main aim is to reduce the incidence of crushing to your ribs and back if a horse steps on you after a fall. They also give some cushioning protection if you land on your back or chest on rocky ground. They will not protect your back from a whiplash-type injury or if you land on your backside or head, for example. This type of fall can cause serious spinal concussion. Landing in a roll is the best protection against this type of injury. Modern body protectors are lightweight and not much larger or heavier than a waistcoat. They must be fitted so that they do not restrict movement and they should not overheat you in hot climates. They should protect and cover vulnerable parts of your body, such as your ribs, stomach and back. Like helmets (and indeed all safety equipment) body protectors are still developing and the future will no doubt bring significant improvements in their design. It is also likely that there will be a legislated increase in their use in some horse sports. In 2005, good body protectors meet ASTM F1937-98, the most stringent US standard, or BETA 2000, the most stringent UK standard. The Australian Racing Board has written a body protector standard that is primarily based on the SATRA standard. SATRA is the largest international research and technology organisation for safety products.

Eye protection

In sunny climates you should wear good-quality plastic sunglasses for handling and riding. Some instructors and riding schools discourage this as it makes eye contact difficult, but you should never be discouraged from wearing sunglasses as they are the

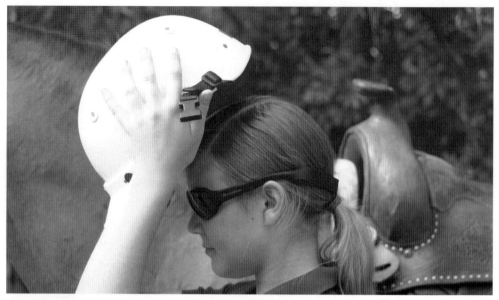
Sunglasses can be secured around the back of your head with a strap

best eye protection possible. The light-coloured sand often used as a surface in riding arenas causes a strong glare, making it difficult to see properly. Employers and instructors must realise that eye protection from sun damage is very important for children and adults alike and that not permitting the use of sunglasses can be an occupational health and safety issue for staff and clients.

Sunglasses can be made more secure by fastening them with a strap behind your head. Prescription glasses can be secured the same way.

Sun protection

Sunscreen (30+ or higher) or preferably total sunblock should be applied to any exposed skin by handlers and riders when outside. It should be kept in an area such as the tack room, for ease of use. Commercial operations should always have high-protection sunscreen/sunblock available for use by clients and staff.

Other protection

Various other items are important for protection in specific equestrian sports, for example face protectors and kneeguards in polo. Some items may be compulsory. Check with the club or official body before competing.

Support

Your lower back can be supported during horse chores and when riding by a neoprene back brace which supports weak core muscles until you have been riding for some time and developed the correct muscles.

People who suffer from knee problems can benefit from neoprene knee supports designed for riding. Knee problems are common; riding may either aggravate or improve the condition depending on the problem.

For many women, a good-quality sports bra is imperative while riding. Women should not ride in underwire bras because the wires can be forced through the ribs and into the chest (lungs and heart) during a fall.

You can buy men's or women's underwear that has been designed especially for riding. These items reduce rubbing and therefore soreness which can spoil a ride. If you plan a long ride try the underwear first, to make sure it is comfortable.

Spurs and whips

Spurs should only be worn by experienced riders who understand their use. You must have an independent seat (see p. 118) in order to use spurs safely and effectively. Spurs should be used as an extension of your foot and can help some riders, for example those who have a loss of muscle in one leg. Whips should be regarded as an extension of your arm and should be used as a training aid rather than to punish the horse. Beginner riders find it difficult to control a whip when they are still developing their balance, so horses that are used for

A neoprene back brace will support your lower back

beginner lessons should be re-educated by experienced riders to respond to the leg cues if they become dead to the leg (you squeeze or kick and nothing happens, a typical response of many riding school horses). The handle of a riding crop (a short whip) should not be placed around the wrist as it can cause a broken wrist in a fall. Cut it off to avoid this happening. Check the rules about whips if you compete as there are variations between horse sports.

What not to wear

There are numerous items that should not be worn when handling or riding horses.

- Jewellery can get caught, for example an earring can get caught in the horse's mane and be ripped out of your ear. Jewellery can cause a multitude of accidents and can complicate an otherwise minor accident.
- Bags, cameras and anything else should never be strung around your neck. These can cause injury, especially if it the item is hard to break (such as a key on a string).
- Keys or a pen in your pocket (hip or chest) can break your bones or stab you in a fall.
- Loose clothing can hook on to the saddle when you're attempting to dismount. Where possible, tuck loose clothing such as t-shirts and shirts into your trousers. Loose flapping rainwear is unsafe because it can frighten a horse and catch on things too easily.
- Riding in shorts is unsafe because they do not give any protection in a fall and the inside of your legs will become sore due to rubbing by the stirrup leathers.
- Crop tops and strappy flimsy tops give no protection against the sun, insect or horse bites or abrasions if you fall.
- Long jeans or pants that drag on the floor. These can trip you when on the ground.

4

Facilities and environment

Facilities should be safe for all users, visitors, clients and horses. Many accidents are caused by unsafe facilities and poor management practices.

When planning the ergonomics of the property, i.e. the immediate environment, take account of factors such as the logistics of safely moving horses and vehicles around the property, where the house, buildings and other horse facilities should be in relation to each other in order to reduce work, and areas that will be used most frequently. A well-planned property is safer and reduces stress and expense through efficient labour and lower running costs.

Planning facilities that have a minimum negative impact on the environment and, if possible, enhance the environment should be a major concern. Land degradation affects not only the environment at large but also the health and well-being of people and horses on the property. Degradation of the environment is caused by poor management practices. Examples of good environmental management include planting vegetation for shade, considering what surfaces to put where to reduce dust and mud, effective manure management and how to reduce the fire risk.

When designing and building a property from scratch you should incorporate good safety and environmental factors into the design. On an established property you may need to make some modifications.

Access

The driveway should not be rutted or unduly steep and you should be able to see both directions along the road when exiting the property. Consider changing the surface materials or altering the access if necessary. Remember that when towing a horse trailer you will need more time to join the road as the vehicle will pull much more slowly than when not towing. Likewise, turning into the property with a horse trailer also requires slow manoeuvring. The width of the driveway depends on the number of people who are going to be using it and the visibility, for example a tree-lined curving driveway will have less visibility than a straight driveway with a low hedge and so it may need to be wider because drivers will not be able to see as far along it. The minimum width should be 3 m to allow for general large vehicles such as trucks delivering hay, and fire trucks or other emergency service vehicles.

On a private property, a single-lane driveway is usually fine as long as visibility is good. However, avoid the situation of a vehicle being stuck in the road unable to enter the property because another vehicle is trying to exit. A commercial property may need more than one driveway, one for residents and one for clients, in order to keep clients separate from residential areas. A commercial property will have much more traffic entering and leaving, so passing places or dual lanes may need to be constructed.

A horse property should always have a closed gate to prevent any loose horses from getting out onto the road. If this gate is at the road end of the driveway, site it so that it is at least the combined length of a car and horse trailer from the road so that a vehicle does not have to remain partly on the road when the driver is opening and closing the gate. This will also be much safer if people will be riding horses in and out of the property. A commercial property wants to have traffic flow easily, so consider siting the gate at the business end of the drive. This has the added advantage of making it easy to see that the gate is always shut!

Cattle grids are not a safe alternative to a gate where horses are concerned, especially near areas where people are riding. Horses can seriously injure themselves, usually breaking a leg, when attempting to cross a grid.

If the property is in a flood-risk area, make sure the driveway will protect against isolation during flooding. For example, the driveway should be constructed so that it avoids low areas. This may mean that the driveway is longer than if it followed a more direct but lower course, but this will be invaluable in a flood.

Users and visitors should be able to move around a property quickly and efficiently in all weathers and without having to go through one paddock to get to another. It is not safe to lead a horse through a paddock that contains other horses. Laneways should ideally link paddocks to the main facilities area for ease of use and safety. On smaller properties it is not usually necessary to be able to drive a vehicle on the laneways, except in an emergency. Laneways should be surfaced if they are used frequently by vehicles. The best location for laneways is on ridgelines where the ground will be dry; if this is not possible, avoid wet areas. Laneways must be wide enough so that horses in paddocks cannot reach people or other horses being led along the laneway. It may be necessary for two horses to be able to pass in the laneway. Taking these factors into account, laneway width should be at least 3 m.

A horse property needs parking areas that are firm and large enough for turning trucks or cars with trailers. These areas should be inaccessible to loose horses. On a commercial property the car parking area should lead directly to the reception/office area so that visitors and clients do not walk through horse areas without supervision.

Fences

Fences on a horse property must be in good condition and appropriate for the area that they are situated in. A good safe fence should keep animals in, intruders out (adults/

children/dogs) and not cause injuries to people and horses. Safe fences need not be overly expensive or difficult to erect. Many horse properties successfully use a variety of fence types because the same type of fence does not necessarily work well in all areas. Fencing is a very expensive commodity so it is important to use the best fence for the job in a given area.

Unsafe fences cause numerous injuries to horses in paddocks, some of which are fatal. Even a fence that is generally regarded as safe, such as post and rail, can cause serious injuries if a horse crashes into it. To reduce the risks of fence injuries, horses should come into contact with fences as little as possible. Below are some ways in which this can be achieved.

- Use electric fencing to keep horses away from wire fences whenever possible.
- Make paddocks as large as possible (taking other factors into consideration) so that horses have minimal contact with fences.
- Avoid putting horses in separate paddocks. This causes them to walk the fence lines (which also cause land compaction) and even to challenge fences in an attempt to get to other horses.
- Consider is there is something on the other side of the fence that will attract a horse, for example better grazing, other horses or feed storage.
- Plant bordering trees so that horses can see where their paddock ends. These trees can serve several purposes, such as habitat for wildlife, windbreaks and firebreaks, and fodder for stock (only certain types of trees).
- Regardless of the type of fencing used for paddocks, it needs to be periodically checked and any problems quickly rectified. This is especially important before horses are released into a new paddock.

The perimeter fence is very important because it is the barrier between animals on the property and the road or neighbours. This fence should be strong and able to hold animals even if they charge into it. For this reason an electric tape or braid fence on its own is inadequate for a perimeter fence because horses can sometimes go through such a

A post and plain wire fence that has an electrified offset wire running through the middle of the post. The electric offset wire reduces the chance that a horse will run into the fence, and even if contact is made, the drilled holes for the wire means that the wire will not pop off the post.

A post and combined plain and sighter wire fence, which is not a suitable perimeter fence due to the wires on the outside of the post. The addition of an electric wire on the inside would make this fence safer.

Annie Minton

A steel fence

A steel pipe fence

fence. Types that are suitable for the perimeter are posts and rails, posts and plain wire, specialised horse mesh, steel or PVC fences. Electric fences in conjunction with some of these materials are also a good option for perimeter fences, and they reduce the chance of a horse running into the fence. Wires in perimeter fences can be run through drilled holes in the posts, as then they will not come off if a horse runs into the post, which can happen if the wires are on the outside of the post. The holes in the posts can allow termites, water and fire to penetrate the post however, reducing its life. Perimeter fences should be frequently checked for wear and tear. Horses can chew wooden rails, making them easier to break and allowing horses escape.

Internal paddock fences can range from permanent constructions, like the perimeter fence, to fences that are electric only.

Even the most expensive and best-looking fences can cause serious injuries to horses so it is far safer if they avoid contact. Electric fencing is excellent for horses because once they have learned to avoid touching it they rarely contact a fence. Electric fencing can also be used in conjunction with a solid fence (as for the perimeter fence). Electric fences can

An electric fence that is suitable as a temporary internal fence

be permanent or temporary; you can buy equipment for either. Temporary electric fences enable horses to be safely fenced while you decide where it would be best to put the permanent fences. It is often a good idea to use an area for a cycle of seasons before erecting permanent fences so that you can see how it behaves in wet weather, for example.

Electric fences should be clearly signed if they are situated in areas where clients/neighbours/bystanders may come into contact with them. In some areas (usually urban) this is a **legal requirement**. Preferably, electric fences should be erected so the current can be switched off when a paddock is not in use and when people enter to catch horses. This involves incorporating switches at specific points in the fence. Electric fence manufacturers can usually provide information on the safe and efficient erection and operation of electric fences, so it is well worth speaking to them before starting.

When introducing horses to electric fences care must be taken that handlers do not get injured. The first time a horse comes into contact with an electric fence its reaction is usually very fast – the most common response is to turn and run. Never hold a horse while it investigates an electric fence! Never stand in the paddock while a loose horse investigates an electric fence. Horses quickly adapt to electric fences. They should be allowed to learn about electric fencing while loose in a paddock on their own, preferably without other horses and definitely without people. Foals should not be introduced to electric fences until they are at least one-week-old.

Note: In hot dry climates there may be restrictions on the use of electric fences during summer months because they can cause fires (when the current arcs between a grass stem and the wire/tape/braid). Once horses are acclimatised to electric fences they may not challenge them for some time even if they are turned off periodically.

The height of permanent paddock fences must be appropriate for the type of horses on the property. Generally, 1.2–1.4 m is the recommended minimum height for paddock

A horse-safe mesh fence

Diamond Mesh Fencing

fences, but some situations require higher fences. The perimeter fence may need to be higher if it is to contain fit young horses, especially if the property fronts a busy main road. When fencing paddocks, another factor to keep in mind is that fences may need to keep dogs or young children out (including visitors', clients', your own and neighbours'). Mesh fences are good for this purpose however they vary in safety for horses. Mesh with gaps large enough for a horse to get a hoof through can be dangerous, for example ringlock fencing developed for cattle and sheep. It may be necessary to fence for several types of stock if you use cross-grazing. The mesh fences, especially those with large mesh, are safer if electric fencing is also used to keep horses away from direct contact. Horse-safe mesh is closely interwoven so that horses' hooves cannot go through it.

The shape of the paddocks will be dictated by the available space and the landscape. Wet and dry areas must be separated so that they can be grazed at the appropriate time of year. Square areas are the most economical to fence but they are unlikely to fit in with designing a property to minimise land degradation if the property is undulating.

Paddocks with rounded corners are safer for horses. Cantering or galloping horses are guided around a rounded corner rather than into it. Never have acute angles in paddocks because horses can be herded into them by other horses. Any dangerous corners in an area that is already fenced can be eliminated easily by fencing across the corner with electric tape or braid. This fenced-off area can then be planted with bushes or trees for extra shade and shelter and wildlife habitat. Rounded corners are easier to maintain because a tractor and harrow or slasher can get up to the edge of a rounded corner.

Gateways

Many injuries are caused by gates and gateways because they are an area where horses tend to congregate and are a high-use area for both people and horses.

People leading horses in and out of a paddock can be injured when horses crowd a gateway. Common scenarios are people getting kicked by horses or trapped between a horse and a gate. Horses tend to congregate in gateways and a cornered horse may try to jump the gate.

Gateways in paddocks should be situated in a high and dry area if possible. They should not be in the corner, to reduce the chances of people or horses being cornered. An electric tape or braid 'yard' with a tape gate can be erected across a corner that contains a gate, or used to form a square around a gateway that is not in a corner. This helps to keep horses away from the main gate and is safer when entering and leaving a paddock with a horse. If a gateway opens inwards only or opens onto a road this strategy is especially valid.

Preventing horses from hanging around gateways will reduce the risks of accidents. As well as using electric tape to keep horses back, avoid having horses standing around waiting to be fed or

A flush-fitting gate with an electric fence warning sign and a fastener that does not protrude into the opening

brought in. This may mean bringing horses in sooner than planned, but if the horses are only standing around in the gateway (rather than grazing) they might as well leave the paddock.

Paddock and yard gates should swing easily, rather than be dragged along the ground. The hinges should not lift if a horse rubs its head or bangs into them. A gate should swing either ways or outwards only. Gates that open only inwards increase the risk of a person or a horse getting trapped or kicked when releasing a horse or taking a horse out of the paddock (see p. 70).

Gates should have easily operated fastenings that can be opened and closed with one hand, such as when leading a horse. Gate fastenings should not protrude, to reduce the chance of people and horses getting hooked on them. Protruding gate fastenings can cause serious injuries to people and horses. Gate fasteners should also be tamper-proof from horses. Trees and bushes around gateways should be kept trimmed for good access.

Gates should be fitted flush with the post on both sides to prevent a horse from trapping its head (when rubbing) or legs (when pawing) between the gate and post. If this happens the horse may pull back and cause serious injury to itself. A gate that does not fit flush can be made safer by fitting an electric tape yard as described earlier, by fitting an electric tape closer across the front of the gate or by adding a well-secured piece of wood between the gate and the post to fill in the gap.

Double gates are not ideal because they are less secure than single gates and are hard to manage if you are trying to prevent both gates from swinging open when leading a horse through them. Double gates can close with a nutcracker action if a horse gets its head or feet caught between them. If double gates are fitted, ensure that they fit flush to one another with a fastener at the top and bottom. Again, the use of an electric 'yard' will keep horses back from the gates making them safer.

Gates should be constructed so that horses cannot get their feet trapped in them. A common injury occurs when horses paw at a gate and trap a foot in it. Gates should be made with very small mesh to prevent feet from going through, or no mesh at all so that a horse can get its foot back out easily. Avoid gates with diagonal bars (if they are not close-meshed) because a horse can trap a foot between the acute angles created by a diagonal bar crossing a horizontal bar.

All paddocks should have a gate that is at least 3 m wide to allow access for emergency services (ie a grass fire in the paddock).

Holding yard gates should open either outwards only or both ways.

Training yard and arena gates should swing both ways and preferably be able to be operated by a mounted rider. The gates should be at least 2.4 m; a width of 3 m is required for vehicles (for maintenance).

Gates should be kept closed when not in use. Gates that open directly onto a road should be padlocked to prevent horses being released onto the road.

Stables, shelters and holding yards

Horses need either a built shelter or natural shelter (trees/bushes) that they can retreat to from the sun, inclement weather and flying insects. When deciding on a site for buildings and holding yards, consider such factors as stock and vehicle access, drainage and wind/fire protection. Stables and other buildings should be located in elevated areas large enough to be used as a holding area in an emergency (flood, fire, high winds etc.).

Stables and holding yards usually need to be at least 50–100 m away from watercourses such as creeks and rivers and at least 2 m above the highest water level for the area. Ask the local council for advice and regulations. If the stables and yards must be near water they must be sited on an impervious layer such as concrete or compacted limestone and bordered with a barrier such as logs so that manure and other debris is prevented from entering the waterway.

Design facilities so that horses can be fed without people having to enter a small space with them. This makes it safer for handlers of all abilities.

Standard stable sizes are 3.6 m (12 ft) square for a horse and 3 m (10 ft) square for a pony. Stables can be smaller or larger than this in certain situations. For example, if a stable is used only to feed the horse in it doesn't have to be as large. On the other hand, if a horse is to be stabled full-time the stable should be at least standard size and preferably larger.

Areas for holding horses (stables, shelters and yards) should not have protrusions or sharp edges that could injure a person or horse. The same goes for any stable fittings such as hayracks and waterers. In stables, doors should either swing out from the stable or should slide. Doors that swing inwards only can trap a person or horse in the doorway. Doors should be kept closed when not in use.

Horse yards should be fenced using safe and strong materials such as steel with no sharp edges, plastic-coated wire or safely constructed (no nails) post and rails. Electric fencing should not be used in narrow yards such as those attached to stables, as a horse cannot move around and relax if it is in such close proximity to electric fencing. Also, it is not safe to be in a small area with a horse that is enclosed by electric fences, for example, when a handler goes to catch the horse. A handler might inadvertently move the horse into the fence, or the horse could knock a handler into the fence.

Training yards and arenas

Ideally, areas that are used for working horses should be situated so that they can be seen from the house and stables, so that other family members or staff can easily check people working horses.

Training yards (such as round yards) and arenas should be constructed for safety. The surface should be even, should drain well and should not be slippery. Fences on smaller

An enclosed arena with the fence placed directly at the edge of the riding area.

An enclosed arena with the fence placed further back

AHSE

training areas such as round yards should be at least waist-high on a mounted rider. This is approximately 1.8 m, depending on the size of horses. Arenas that are used to teach beginner riders should be fenced (rather than open) and the fence should be of high enough to discourage horses from jumping it. This depends on the size of horses/ponies using the arena, but 1.4 m is usually a suitable height. Arena fences should not consist of a single rail that a ridden pony might go under.

Never use steel stakes (star pickets) for fencing riding and training areas because of the risk of impalement. Electric fencing (either live or not connected) and barbed wire are also highly unsafe. In training yards, high solid fences that you can't quickly escape through are not as safe as fences that you can get through if necessary.

All fences on training yards and arenas should be constructed so that there are no protrusions on the rider's side where knees could be caught or clothes snagged. Particularly beware of protruding nails on rails. Wooden fences should be checked for slivers of wood. All fences should be periodically checked and any problems quickly rectified.

An arena that is used by a number of people should have a list of rules posted at entranceways to reduce the chances of collision, kicking etc. (see p. 116).

A steel portable round yard. This type of round yard can be used as a temporary measure or on a permanant basis. Never tie a horse to such a yard as it can move.

Air quality in stables

A factor that is not usually taken into consideration is the air quality of stables and the dangers of breathing airborne pollutants, for both people and horses. Airborne pollutants in stables are a combination of gases, bacterial and fungal toxins and spores, mites, pollen, feed particles and animal components. These come from sources such as urine, manure, feed (hay and other feeds), bedding and animal skin, respiration, outside dust blowing into the building (soil and manure) etc.

These pollutants are always present in a stable, but when a stable is being cleaned out a much larger number of pollutants circulate due to the disturbance of the bedding.

These pollutants are inhaled by whoever is cleaning the stable (mucking out) and by any nearby horses. You can reduce the dangers of airborne pollutants in stables.

- If you are building, build stables that are at least the minimum standard size and preferably larger, and that have good ventilation. Use lining materials that trap less dust, for example smooth sheet steel, smooth concrete or planed timber trap less dust than rough finished materials such as rough sawn timber and rough finished concrete.
- Periodically power-hose the stables from top to bottom to get rid of cobwebs (which collect dust and are a fire hazard), dust and other pollutants. This can be done when there is plenty of available rainwater; if water is not available the building can be swept with cobweb brushes.
- When mucking out, wear a mask and remove any horses from the stables.
- Minimise the time that horses have to spend inside. Horses are more sensitive to airborne pollutants than other animals and are generally exposed for much longer than other domestic animals. Yards attached to stables allow horses to access fresh air.
- Store hay away from the stable building. Hay can be soaked or steamed just prior to feeding to reduce fungal spore inhalation. Use only good-quality hay. Dust and fungal spores originate mainly from hay and bedding. Poor-quality hay is dustier and has more fungal spores than good-quality hay.
- Use absorbent low-dust bedding. The main types of bedding for stables are straw (wheat, barley or oat), hardwood or softwood shavings, or sawdust. Less common alternatives include shredded paper and commercial beddings such as cleaned, dust-free shavings or compressed wood pellets; these are more expensive. Straw is usually dustier and has more fungal spores than wood shavings and sawdust, and is less absorbent. Shredded paper has been used with some success, as it is absorbent and not dusty. Commercially produced beddings have the lowest dust rating but are more expensive.
- Use rubber mats on the floors to reduce the amount of bedding required. A small amount of commercial low-dust bedding can then be used in one small area to soak up liquids. This is cost-effective due to the small amount required.
- Keep the area immediately outside the stables as dust-free as possible. A well-watered grassed area is less dusty than bare earth.

Recommended further reading

Avery A (1997). *Pastures for horses: a winning resource*. RIRDC, Sydney.

Cargill C (1999). *Reducing dust in horse sables and transporters*. RIRDC, Canberra. This report can be viewed and downloaded from the Rural Industries Research & Developments Corporation website: www.rirdc.gov.au.

Foyel J (1994). *Hoofprints: a manual for horse property management*. Dept of Primary Industries, SA.

Myers J (2005). *Managing horses on small properties*. Landlinks Press, Melbourne.

Fire and electrical safety

Stables are very prone to fire because of the materials within. The situation of a horse property is often prone to bushfires and grass fires. The risks of stable fires can be reduced with good management and stables can be constructed of low fire-risk materials such as steel or concrete. Other things can be done to reduce the risk of stable fires.

- Do not store hay in the stable building.
- Keep stables free from cobwebs. When cobwebs become covered in dust they lower the flashpoint.
- Do not store flammable material such as petrol, kerosene or paint in or near the stable area.
- Do not burn refuse close to the stables.
- Keep electrical extension leads off the ground and put them away straight after use. Never lead a horse over an electrical extension lead.
- Light switches and power points should be out of horses' reach.
- Electrical equipment should be tested and inspected by a qualified electrician.
- Electric cables should be protected from chewing by rodents and horses. Enclose cables in heavy-duty conduit.
- Know where the electrical master switch is.
- Never padlock stable doors.
- Have regular fire drills and decide where to put the horses in the event of a fire.

Ideally, equip stables with the following.

- An alarm system that is regularly tested.
- Suitable fire extinguishers, also regularly tested.
- A hose permanently attached to a tap, with sufficient length and pressure to reach all buildings.
- Prominent 'no smoking' signs.
- Heavy-duty smoke alarms, which are cleaned regularly.
- A sprinkler system, bearing in mind that this may be expensive.

Stables should have fire-fighting equipment, such as a fire hose and fire extinguisher

Check state requirements on fire protection. These requirements vary depending on the building code and the use of the buildings.

In the event of a fire, you should do the following.

- Call the fire brigade, give all the details (including the emergency services street number) and wait to check that they have understood before hanging up.
- Check that people are safe and out of the building.
- Only if safe to do so, lead the horses out and put them somewhere where they cannot run back in or impede firefighters.
- Aim to minimise the entry of oxygen by keeping building doors closed.
- Make sure the entrance is clear and unlocked for firefighters.
- Don't go into a building if the fire is dangerous and don't go back into a burning building once you have come out.
- Fight fire using hoses, wet feed sacks, fire extinguishers, sand, shovels etc.
- Once the fire is under control, call the vet if any horses are injured. Hose the burns gently until the vet arrives.
- Send someone to meet the fire service and direct them to the emergency.

In some areas grass and bush fires are a common problem. Plan to make the property as fire-resistant as possible by good placement of buildings and shelter belts and routine maintenance. Good property design and carefully worked-out plans are essential to minimise the impact of fire. Keep in mind that local councils may require a special permit to build within a fire-prone area.

Talk to neighbours or the local fire service about local fire risks.

- Is fire likely to occur in the area?
- How often does it tend to occur?
- Is there a warning period?
- Does it follow a predictable pattern?
- Which direction do the hot dry winds tend to come from? Fire usually comes from a specific direction, but be prepared for the chance of it coming from a different direction.
- Is the property part of a community fire-reporting network?

When planning a property from the beginning, implement fire-safe features. An established property may require some alterations to make it safer.

Think of the property in terms of zones. The inner zone includes any dwellings, stables, yards and barns (the most valuable parts of the property), surrounded by a buffer zone of about 20–30 m. The inner zone and buffer zone must be kept clear of debris such as dead leaves, fallen branches, rubbish, firewood etc. The outer zone is the pastures and wooded/bush areas. Draw up a plan and identify the different zones. Your plan should include neighbouring properties, as the fire could come from their direction.

Possible problems should be identified and counteracted. Fire-risk problems include areas with long grass, stacks of firewood, loose hay, chemicals and certain trees. Remember that fire travels faster uphill. Pluses include dams, swimming pools, drives and laneways (which can be used for firebreaks, evacuation routes and for firefighting access), gravelled areas, short grass (perennial pasture is both good summer feed and fire-resistant), ploughed land, deciduous orchards, vegetable gardens, arenas and windbreaks of fire-retardant trees. Plan to place firebreaks in the directions that fire is most likely to come from and reduce any fire-risk problems in these areas. Property owners might be required by law to create a firebreak around the boundary of their property. This could be in the form of slashing with a tractor or plowing. Check with your local council or fire services.

In the inner and buffer zones the grass must be kept short and green if possible. Any combustible material should be removed. The ground under all trees must be kept clear. Leaves should be cleared up regularly. Laneways and/or firebreaks should be grazed or mowed regularly to keep them short. Any long grass should be mowed or slashed/whipper-snipped.

If the property has water tanks or large water bodies such as dams or lakes it is a good idea to get a petrol-driven pump for firefighting, because the power is often the first thing to go. The pump can be situated next to the power pump with a connection to the same system. A sprinkler system around the inner zone will be invaluable in a fire. Make sure all connections are metal not plastic, as plastic will melt in a fire.

Consider having the local fire services visit your property to evaluate the facilities and demonstrate how to attack fires. A fire evacuation plan is a must in fire-prone areas, and the fire service can advise on this also.

Consider purchasing an emergency management manual that will assist you with the management of any emergency on the property. Talk to your local fire service about this.

Recommended further reading
The website of The Equine Centre (Melbourne) has articles on horses and bushfires:
www.equinecentre.com.au.

5

Safe horse handling

As many as 20% of injuries involving horses occur when people are handling and near them on the ground, not riding (Sport & Recreation Vic.). Common injuries when handling horses are from biting, kicking, headbutting, trampling and getting squashed between a horse and a solid object. Most of the injuries are not deliberate attempts by the horse to harm a person, but this does not make the injury less severe. Injuries often occur simply because a person inadvertently put themselves in a high-risk situation.

When handling horses you must have at least a basic understanding of horse characteristics (see p. 4) and how these can affect safety.

Halters and leadropes

The most commonly used tools for handling horses are a halter and leadrope. A halter can be made of leather, synthetic materials or rope. Rope halters are made from one long piece of knotted rope. Leather or synthetic halters with adjustable nosebands are useful as there are occasions when the noseband needs slackening (inspecting the horse's mouth or rasping its teeth) or undoing (bridle being put on).

Halters should be made from good-quality strong materials, and leadropes should be at least 1.5 m long with a heavy-duty (spring-loaded) clip at the end.

There are variations to the standard clip on a leadrope, such as a panic snap which

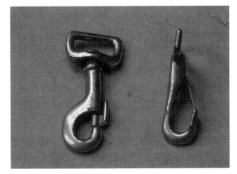

The clip on the right should not be used as part of a leadrope as it can get caught in the horse's lip and cause injury if the horse pulls back. The clip on the left is a standard 'dog' clip and is usually safer to use on a leadrope

comes apart if a horse pulls back quickly. These can be useful in some situations, such as when cross-tying a horse, but they are inappropriate if the horse can escape onto a road, for example, if it breaks free.

Putting on a halter

As with many horse-related activities, there is more than one way to do this. The main considerations are to make sure that the horse doesn't walk off as you are attempting to

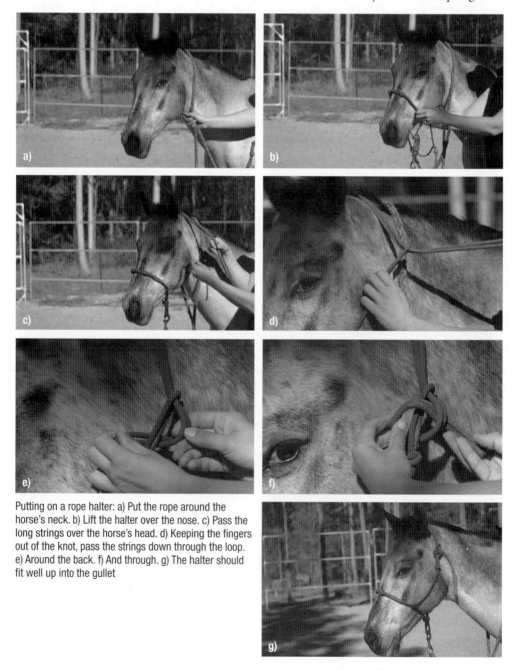

Putting on a rope halter: a) Put the rope around the horse's neck. b) Lift the halter over the nose. c) Pass the long strings over the horse's head. d) Keeping the fingers out of the knot, pass the strings down through the loop. e) Around the back. f) And through. g) The halter should fit well up into the gullet

fasten the halter. Be careful when handling sensitive areas such as the nose and ears.

Never leave a halter on a loose horse, because it can catch on projections. A common accident occurs when a horse catches its headgear on the door catch in a stable and then pulls back. A horse can catch a back hoof in a halter if it attempts to scratch an ear while wearing one, and in paddocks a horse can catch a halter on fences or branches.

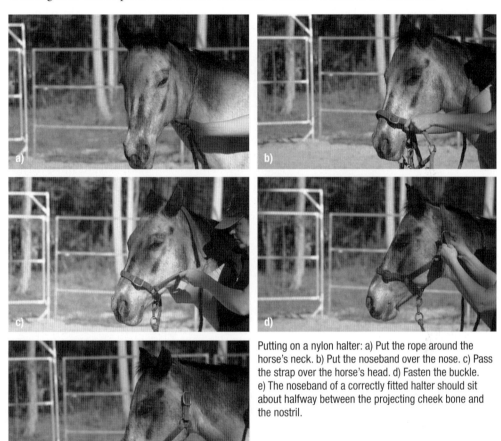

Putting on a nylon halter: a) Put the rope around the horse's neck. b) Put the noseband over the nose. c) Pass the strap over the horse's head. d) Fasten the buckle. e) The noseband of a correctly fitted halter should sit about halfway between the projecting cheek bone and the nostril.

Safety procedures in horse areas

The areas in which horses are kept and handled should be kept tidy at all times to reduce the risk of accidents caused by things left lying around. Every item should have a home and be put away as soon as it is no longer needed. Cleaning tools should be hung in a safe spot where they cannot trip someone. Hoses and cables should never be left on the ground. There should be no protrusions that could injure a person or a horse.

A child should never be left in a pram in a horse area. The child is in a very dangerous position and a horse will not initially assume that a pram is harmless. A child in a pram is at a very dangerous height if a horse kicks its pram. It is unable to move away, and can also be knocked over and trampled.

Before mucking out a yard or stable the horse should be secured or removed. Removing the horse is preferable as it is safer (it is impossible to concentrate on what the horse is doing at the same time as cleaning) and it reduces the amount of airborne pollutants that the horse is forced to inhale (see p. 36).

Keep aisles and breezeways in stables clear and tidy. Even in stables where many people share the same spaces, tack trunks and other paraphernalia should be stored in tack storage areas rather than outside individual stables.

Annie Minton

All tools should be hung safely

Any chemicals such as weedkillers and worming pastes should be stored in a safe place such as a lockable cupboard or room.

Halters should be hung up after use, in an area that is easy to reach quickly in an emergency. Either hang all halters in the entranceway to a stable block or hang a halter for each horse directly outside individual stables. Any hooks in a horse area should face into the wall to reduce the chance of horses getting caught on them.

A halter and lead that has been removed from a horse that is being exercised should be fastened up off the ground until the horse returns. A passing person or horse can be tripped by a hanging halter.

Paddocks should be safe for both people and horses to use. Make sure that there are no farm implements there, such as harrows, which quickly get covered in grass and become hazards for loose horses. Loose fence wire, holes from rabbits or tunnel erosion, stakes sticking out of the ground, projections on fences and low branches are all potential causes of accidents.

Separate people and horse first aid kits are required in horse areas, plus emergency numbers for doctors and vets. These should be displayed clearly. Have a second-choice vet phone number as well, in case your first choice cannot be contacted. Consider doing a first aid course; the knowledge gained is useful for dealing with horse injuries as well human injuries and emergencies. First aid kits for people can be purchased relatively inexpensively at various outlets such as chemists. A first aid kit for horses can be purchased as a kit or made up from individual items. Talk with your vet about the items required for a horse first aid kit. Both kits should be easily accessible, clearly marked and distinguished from each other.

A commercial stable should have signs posted at key points such as stable entrances warning people of the dangers and outlining the safety rules that visitors and clients must adhere to. Even a non-commercial stable should consider posting signs for when people visit the facilities. For example, 'no smoking' or 'helmets must be worn' signs give a clear message. Risk-warning signs explaining that horses can be dangerous are a good investment!

Recommended further reading

Managing health and safety in the racing industry (Qld). This is a booklet on workplace health and safety issues related to racing (much of the information is also relevant to other sectors of the horse industry). It can be viewed on the links page of the Australian Horse Industry Council website: www.horsecouncil.org

Basic horse management requirements

Horses have basic management requirements if they are to perform safely. The care of horses is a complex subject and will be discussed here mainly in the context of the safety of handlers and riders. In addition to a constant supply of clean fresh water (up to 70 L of water a day) a horse also needs correct feeding (see p. 60), regular hoof care (see p. 58), regular dental care (see p. 60), companionship (see p. 8), shade/shelter (see p. 34), exercise (see p. 121), grooming (see p. 55), parasite protection (see p. 64) and insect protection (see p. 56). Appropriate rugging is required in some situations (see p. 62) and inoculations may be necessary depending on the disease risk in a particular situation (see p. 65).

Recommended further reading

Huntington P, Myers J & Owens L (2004). *Horse sense: the guide to horse care in Australia and New Zealand*, 2nd edn. Landlinks Press, Melbourne.

The website of The Equine Centre (Melbourne) has lots of useful articles on health and welfare of horses: www.equinecentre.com.au

The Commonwealth Serum Laboratories website (www.csl.com.au) has more information on vaccinations.

The Rural Industries Research & Development Corporation (RIRDC) website (www.rirdc.gov.au) summarises local research on nutrition and general health relevant to Australian conditions.

Catching a horse

Catching a horse is not something that a complete beginner should be allowed or encouraged to do until they have a basic understanding of horse characteristics and have practised putting a halter on and taking it off an already secured horse. They should then try catching a horse in a yard or stable. You must be competent and confident before you attempt to catch a horse in a paddock that contains other horses. Being among a group of loose horses is no place for inexperienced handlers, small children or anyone other than competent horse handlers.

The procedure for catching a horse is the same whether it is in a stable, yard or paddock. The only difference is that in a larger area you can walk around to approach the horse from the correct angle and you can't do this in a small area such as a stable or a small yard.

In a yard or stable

You should never enter a small space (such as a yard or stable) if the horse is facing away from you as this will mean that you are approaching the horse from an unsafe angle. You could startle the horse and be kicked simply because the horse did not see you approaching (see p. 4). The horse should be encouraged to come to you. This can be done by calling it over (some horses will come to call) or by clicking your tongue so that the horse moves. A constant tongue-clicking noise irritates a horse, which moves away from it. In a small area this means that the horse eventually has to turn towards you. When the horse turns to face you, you should stop the clicking. The horse is rewarded for turning to face you, by stopping the irritating noise. If this is done often enough the horse learns to face you when it hears the click. You can then approach and touch the horse and put the halter on.

Remember when touching a horse that they are sensitive (see p. 4). This sensitivity means that it is safer (and more comfortable for the horse) to touch them with a firm flat hand rather than with very light fingertips. Avoid patting horses, especially on the head because the head is largely hollow so pats are unpleasant. Scratching with the fingertips is better because it is closer to what horses do when they touch each other when they nibble and rub with the muzzle. When you approach a horse to catch it, walk towards the neck and scratch the neck before putting the halter on. Always have the halter ready to put on before approaching the horse. This will minimise the amount of time standing next to the horse unhaltered, and so minimise the risk of the horse escaping or injuring you.

In a paddock

If the horse is in a paddock with other horses, catching it is more risky. The dangers involved with catching a horse in a group is that you might end up in the middle of a group of loose horses while trying to catch the one you want, and as you are leading your horse out of the area another horse may challenge the one being led. This situation is more likely if the horse that is being led is low in the group's pecking order. As stated

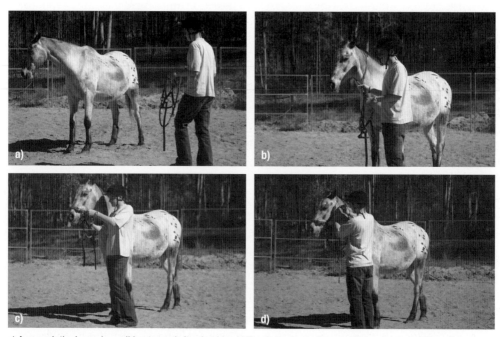

a) Approach the horse by walking towards its shoulder. b) Touch the horse then put the rope around its neck.
c) Lift the halter over the nose. d) Fasten the strap.

earlier, a person that is not competent and confident should not attempt to catch a horse in this situation.

The procedure for catching a horse in a paddock is the same as for in a small space, with the advantage that in a larger area you can walk around to approach the horse from the correct angle. Again, always have the halter ready to put on the horse before approaching it. This will minimise the amount of time standing next to the horse unhaltered and so minimise the risk of the horse escaping or injuring you. When approaching a horse to catch it, you should approach at an angle between the head and the shoulder. Never approach straight towards the head from the front or from behind the horse as these are the horse's blind spots and it may get startled and kick. These approaches also tend to drive the horse away from you – not the desired result! You should speak as you approach and look for an ear movement towards you that will signify that the horse is aware of you (see p. 4). You should walk at a steady pace and keep your arms as still as possible. Any sudden movements will send the horse away.

As you get close you can reach out and rub or scratch the horse's shoulder or neck, then put the rope around the neck and put the halter on.

If your horse is in a group of horses, lead the caught horse around (not through) them even if this means initially moving away from the gate. You should watch for other horses coming up behind while leading the horse towards and out of the gate. The other horses can crowd around and cause the led horse to crush you against the gate, or you may be kicked if the horses start to fight each other. Also, the other horses can rush past you and out the gate if they are allowed to gather there. Watch the body language of both the led and the loose horses for warnings of possible interactions (see p. 33).

Problems with catching

Some horses are hard to catch, which can be extremely frustrating. Some reasons that the horse may be hard to catch include:

- If there is an abundance of good grass the horse may not want or need to be caught.
- The horse may have been handled roughly in the past after being caught.

Initially the horse will need to be separated from the group and put in a smaller area (such as a yard or stable) until it has learned to be caught.

There are many theories about how to teach a horse to be good to catch. One of the ways is to teach the horse to face up as described earlier. The horse can then be put in a larger area such as a round yard for you to see you can easily catch it in there. If not, it needs to go back to the smaller area for more practice and so on. Eventually the horse will be able to be caught in the larger area and then it can progress to a small paddock. Eventually the horse can be released back into a group of horses.

Many theories suggest using feed to catch a horse, but horse people have very divided opinions about whether this should be done. It is never safe to take feed into a paddock or large yard that contains more than one horse. If the horse is on its own you can use feed as a reward to teach it to be caught, if you follow safe procedures for hand feeding (see p. 62).

Some horses are hard to catch

Leading a horse

It is customary to lead a horse by walking on the horse's left (near) side between its head and shoulder. This is the side that most horses are taught to lead from, but there are times when a horse may need to be led from the other side. You should practise this once you are proficient at leading on the left side, so you and the horse are comfortable with leading and being led from both sides.

When on the left side, hold the leadrope by placing your right hand just after but not on the clip. A clip can catch fingers and injure them if the horse moves its head suddenly. Watch children carefully in this situation as it is very easy for their smaller fingers to slip into a clip. Many people have lost fingers in this way. The rest of the rope is held in your left hand. If the rope hangs down to the ground the excess should be folded across the palm of your left hand. The rope should never be wrapped around your hand or any other part of your body. Again, watch children carefully as they tend to play with the rope even while they are leading a horse, and can wrap it around themselves.

A horse can also be led in a bridle. In this case, hold both reins behind the bit with your right hand and hold the excess length of reins in your left hand.

The horse should be made to walk with its head facing straight forward. You should also look forward to where you are going, allowing your hand to move with the natural up-and-down movement of the horse's head as it walks. If the head is held tight its natural movement is restricted and the horse may start to resist. You should not let yourself fall back level with or behind the shoulder. At this point a horse can easily push you sideways if it moves across suddenly (shies). Also, if you are too far back an exuberant horse can kick sideways and catch you with a back foot.

When turning the horse, push the horse around the turn (away from you) rather than pulling the horse around the turn (towards you). This reduces the chances of the horse standing on you or knocking you over as it turns.

When approaching narrow openings such as gateways and doorways the horse should be stopped so that you can go through first. The horse should not be turned sharply until all of its body has passed through the opening. Gateways and doorways should always be opened fully and preferably away from the direction of travel so that the horse doesn't bang its hips as it goes

A horse being led correctly

through or get trapped in the opening. If a horse bangs the hips it may become unwilling to go quietly through narrow openings in the future. This leads to the horse rushing through narrow openings, which is dangerous because you can be knocked over.

Problems with leading

Horses that are difficult to lead should only be handled by people who are competent and confident. Never lead a horse on the roads or in open spaces that empty onto a road if the horse is not quiet in traffic or is afraid of dogs etc.

When being led, the horse's attention must be on the person leading it. You should not let a horse that you are leading or holding 'meet and greet' with another horse, eat grass or feed. If the horse becomes unruly in any way this means that the horse is no longer focusing on you, and you must get its attention back as soon as possible. You can do this by vibrating the lead or, if necessary, using sharp tugs rather than a steady pull. A steady pull on a lead just gives the horse something to lean against. You can put your elbow into the horse's neck so that the horse has to bend its head towards you (giving you the advantage) or you can ask the horse to lower its head (see p. 125).

If a horse rears while being led you must keep out of the way of the front feet as they paw the air to balance the horse. Pulling down on the head can make the horse go higher and even fall over.

If something scary is approaching the horse from the right side when you are on the left side, the horse should be turned towards the perceived danger so that you don't get knocked sideways if the horse shies away. The horse should be made to stand still while the danger passes.

Sometimes it is safer to let a horse go than to hang onto it. You should release the horse if you would otherwise be dragged. The horse can always be caught again. Many people have been seriously injured or killed because they have not let go when they should, thinking that hanging on is more important. Also, a horse can panic further at the unfamiliar sight and feel of a person being dragged. A horse may think that the person being dragged on the end of a leadrope is chasing it, so it panics further to get away from this scary object. It may kick as well to try to rid itself of the danger, with dire consequences for the person being dragged.

If a horse won't move you can't pull it somewhere that it doesn't want to go. If the horse won't move forward you should stand between its head and shoulder and flick the

If the horse will not lead forward you should flick the horse with a whip held in your left hand

lead in your left hand behind your body at the horse's flank. If it is likely that the horse is going to resist in this way, it is a good idea to carry a dressage whip so that you can flick it with that rather than the rope.

Tying up a horse

Tying up a horse can be very dangerous unless you follow these simple guidelines. Even a correctly tied horse can get into trouble and panic, but there is usually less damage if the horse is tied properly. Common incidents and accidents that occur when horses are tied include those listed here.

- Horses that are tied with too long a rope can get a leg over the rope or the rope can pass over the top of the head after they lower it then bring it up again. Both these situations can cause the horse to panic.
- Horses that are tied using poor-quality or the wrong equipment (such as by the bridle reins) can pull back and break free. In the process it can injure people and/ or itself.
- Horses that are tied too low can pull up and back against the pressure. In addition to the pull on the horse's head from low down, which can frighten a horse, the horse can exert more pressure when pulling upwards than when pulling at head-height or above.
- Horses that are tied to an unsafe object such as the towball hitch of a parked vehicle, the wires of a wire fence, a gate that can be lifted off its hinges or an unsecured horse trailer can pull back and cause serious injury to people and/or themselves.

If a horse cannot be tied safely it should be held instead.

Many horses are not taught to tie up properly in the initial stages of training, which will lead to problems later on. Once a horse develops the habit of panicking when tied ('pulling back') it is very difficult to eradicate the behaviour. There are two schools of thought when it comes to tying up a horse.

- The horse should be tied in a way that allows it to break free if necessary. The advantages of this method are that the horse will break free if it pulls back rather than panic further and either break the tying equipment or the object that the horse is tied to. The disadvantages are that the horse may learn to break free and then do it whenever tied; it may escape and injure people.
- The horse should be tied so that it cannot break free. The advantages of this method are that the horse cannot break free and it learns that it must stand still when tied. The disadvantages are that the horse can injure people who are too close, or itself, if it panics.

Both methods have merit and you must use your own judgement to decide which the best method. In reality it depends on the situation at the time, which includes the quality of the equipment, the solidity of the object that the horse is to be tied to and the individual horse. For example, if the second method were used to tie an inexperienced horse or one that pulls back to a post that is not truly solid, the horse may pull the post out of the ground. You should adopt the method that is correct for the time and the place rather than have a hard-and-fast rule.

There are general rules for safely tying up a horse.

- Always use good-quality equipment that fits well and is strong. A good-quality halter (leather or nylon) and rope is best. Don't use chains as they cannot be cut in an emergency. Round rope is better than a flat lead, which will not loosen easily if pulled tight. Never tie up with the bridle reins as the bit can break the horse's jaw or smash its teeth if it pulls back. Rope halters made from narrow twine are also dangerous as they can damage the area behind the poll if the horse pulls back.
- Select an area that has a non-slip surface and is free from clutter. Make sure there are no projections that could cause injuries, such as nails sticking out of a post. This area should not have any loose horses, and tied horses should not be able to reach each other to kick.
- Carefully select the object to tie the horse to. A free-standing solid wooden post or one that is part of a post and rail fence is suitable. It must be well set into the ground. Posts that are part of wire fences are not safe because a horse can put a foot in the fence and panic. Beware of hitching rails if they are not strong; they can come off if the horse pulls back. Never tie a horse to an object that the horse can move. Remember that a horse is very strong, especially when it is panicking. Don't tie a horse to a horse trailer unless the trailer is hitched to a car or truck. Never tie up to a single horse trailer even if it is attached to a vehicle, as a horse can pull it over.
- Tie the horse level with its wither height or higher. The rope should not be long enough for the horse to reach the ground when tied; about 50 cm is enough between the head and the post.

- Use a quick-release knot so that the horse can be released quickly in an emergency. Always have a knife handy in case the horse has to be cut free. Between the rope and the post, use a piece of twine that will break if there is any doubt about the horse's behaviour when tied.
- Be careful to keep your fingers out of the loops when doing knots. It is very easy to lose a finger in this situation if a horse pulls back.
- Don't leave a horse unattended when tied. Horses can get into trouble very quickly when tied, especially at a show or event where there is a lot happening that is out of the ordinary to the horse.

Tying a horse in cross-ties can be dangerous because the horse can pull back and fall and end up hanging from two ropes instead of one. Only cross-tie on a non-slip surface

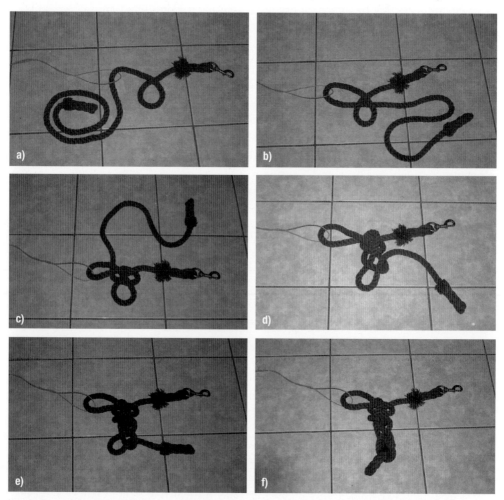

How to tie a quick release knot. There are many types of quick release knot; this knot is called a stockman's bowline. a) Thread the rope through the twine and make a loop in the section between the horse's head and the twine. b) Wrap the loose end of the rope around this loop. c) Pass the rope around and over. d) And down through the loop. e) If the rope is long, one or more loops can be 'daisy chained' from the first loop. f) Finish by passing the free end back through the last chain so that the horse cannot release itself by pulling on the end of the rope. When tying knots keep the fingers out of any loops.

(not smooth concrete), with ropes that are breakable, have panic snaps or have single twine between the rope and the posts. The ropes should be loose but not too long, nor too tight. They should be tied at just above head level. This method is commonly used for stallions in company to restrict their ability to move around and bite any handlers.

A horse safely tied up. The horse cannot lower its head to the ground, as it is tied at a good height, with a quick release knot to a suitable post.

A horse in cross ties standing on rubber mats

Problems with tying

The most common problem with tying horses is 'pulling back'. A horse that pulls back is a horse that panics when tied up. The more the horse does this (as with all behaviours) the more entrenched the behaviour becomes and the harder it is to fix.

Horses that pull back will do so until something breaks or they end up on the ground. A horse can also crash over backwards if the tie breaks, and land on a person: it can also swing its head from side to side vigorously in an effort to get loose, so it is unsafe to approach a horse that is pulling back. This behaviour is very dangerous for all concerned. Every time a horse pulls back and breaks free it is reinforcing the behaviour. If the behaviour is entrenched the horse should never be tied (which is limiting) or should be sent to a professional for retraining – unless it is handled by experienced horse handlers using correct equipment and facilities the horse will become worse rather than better. Even having the horse retrained by a professional does not guarantee success because panic reactions in horses, once entrenched, are very difficult to eradicate.

The reason that some horses pull back is because they were not taught to yield the head and therefore to tie up properly in the first place. A horse that yields all parts of its body to pressure rather than fights against pressure will lead anywhere, will tie up properly and so on. The solution to preventing the problem in the first place, and to solving the problem, lies in teaching the horse to yield rather than pull back from pressure on the halter (see p. 125).

Moving around horses

You should always let a horse know where you are by speaking as you move around it. It doesn't matter what you say, but the sound of your voice will enable the horse to

Stay close to the horse, touch it and speak as you walk around the back so that it knows where you are

locate you when you're in its blind spots (see p. 4). Many kicks occur because the horse was startled and kicked in defence.

When moving around horses you should always stay alert and observant. You should stay either very close to the horse or at least 3–4 m away, preferably more, remembering that a horse can run backwards to kick. 'Very close' means within an arm's length, because at this distance a kick is more a shove. Also, if you are touching the horse you will feel the muscles tense or the weight shift as the horse prepares to kick. Horses that are kicking each other either lean towards each other or spring away to get some distance between them. They know from experience that the endpoint of a kick is the most dangerous place to be because a kick is at its most damaging at its extremity. At this distance the leg is travelling at full speed and the hoof does the most damage.

If you need to cross from one side of the horse to the other you shouldn't go under the neck and rope if the horse is tied up. You may be crushed between the horse and the

Always stand to the side rather than behind the horse when brushing the tail

object that it is tied to if something causes the horse to jump forward, or the horse can strike forward with a front leg. Don't forget that a horse can't see an object that is directly in front of the face and under the head. For this reason, speak before touching a horse on the head. Depending on your hand's direction of approach, the horse may not be able to see it coming.

When moving behind a tied horse to cross to the other side you should stay close and touch the horse on the rump as you pass behind. Remember; speaking to the horse lets the horse know where you are. You should move at a steady speed without rushing or lingering. Never walk behind a horse that is eating, because then its attention is not on you. Never stand directly behind a horse. Tasks such as washing or brushing the tail and taking the temperature can be done from the side.

Avoid being in a small space with a horse. Horses have been known to crush a person against a wall when panicking. Aim to have space around you and the horse, especially if the horse is young or nervous. The area should be secure, of course, but not so small that you are forced into a tight space with one or more horses. Examples are when loading and unloading horses into transport and when using a horse crush. Never squeeze between a horse and a solid wall or partition.

Children should always be supervised around horses. Small children can get excited at seeing horses but they don't know the dangers and horses don't always recognise children simply as small people and can get frightened or defensive if they are not used to them. Added to this, children tend to be noisier than adults. Children must be taught the correct way to approach horses and must be shown how to move safely around them.

Grooming

When grooming, as with all handling, move carefully around the horse following safe procedures (see p. 53).

Horses that live outdoors without rugs need minimal grooming because they take care of their own skin by mutual grooming (two horses groom each other) and they roll in mud and dust to slough off dead skin. The skin also secretes oils that help to keep the horse waterproof and warm. Rugged horses (irrespective of whether they live indoors or outdoors) need regular grooming to get rid of the build-up of dead skin that accumulates under the rug. All horses need to be groomed before being tacked up for riding.

Grooming ranges from a basic brushing to remove any dirt, grit and old sweat marks before riding to a full groom for the purpose of massaging the horse or turning it out for

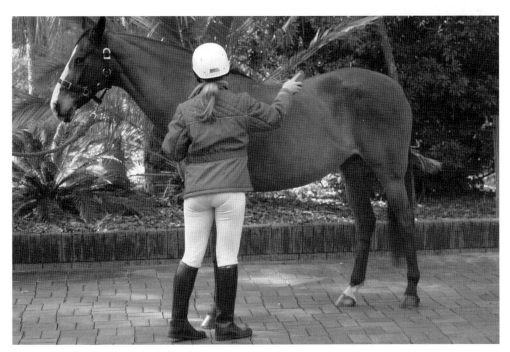

A horse should be groomed before being ridden

a competition. From a safety aspect, only the basic brushing is necessary before the horse is ridden. The areas that are particularly important are where the gear sits on the horse, namely the head area and the saddle area including the girth line. These areas, and the rest of the body if necessary, should be brushed in the direction of the hair growth with a firm bristled brush. Some horses are quite sensitive to being brushed so a softer brush or a glove with rubber pimples can used to break up any dirt. A bare hand should then be run over these areas to check for any lumps of grit etc. that the brush/glove may have missed. The hooves should be picked up and picked out and checked for loose shoes at the same time (see p. 58).

Applying insect repellents

After grooming, insect repellent can be applied if insects are a particular nuisance or the horse is particularly sensitive to them. Horses should be protected from insects as they cause various problems including simply being a nuisance, causing horses to kick out (especially forward towards the belly with a back leg), causing horses to panic and run, and spreading disease. Spray-on repellents should be used with care. Horses that are not used to sprays can panic at the hissing sound and pull back. Always untie and hold a horse to spray it. Never spray around the head area; instead, spray a cloth and wipe it on the head.

Handling the legs and hooves

Handling a horse's legs and hooves requires agility (to move out of the way if necessary) and a good strong back (because it requires bending and supporting a weight). People with weak backs or back problems should get the help of someone else if required. Young children should only attempt to pick up the hooves of well-trained ponies and, as with all horse activities, should be supervised to make sure that they do it safely. In particular,

a) Stand at the side of the horse and b) run your hand from the shoulder down the leg. c) Grasp around the pastern, lean on the horse and lift the leg

watch that children don't get underneath the horse accidentally or on purpose. A horse cannot see underneath the body and it may spook or simply kick forward with a back leg at a fly, even an imaginary fly. You should never kneel or sit on the ground near a horse when attending to its legs. You should instead bend over so that you can move away quickly if necessary.

Inexperienced horse handlers and inexperienced horses are a particularly bad combination when it comes to picking up the hooves. Inexperienced horse handlers need to practise their technique on a horse that is happy about its hooves being picked up. An inexperienced horse very quickly learns to kick out at attempts to pick up its hooves if it is not handled expertly in the beginning.

To pick up a foreleg, stand at the side of the horse with your shoulder next to the horse's shoulder (the area above the foreleg), facing towards the back of the horse. Your nearest hand

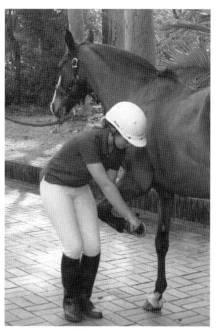

Cleaning out the hoof

is placed flat on the horse's shoulder then run down the back of the foreleg to the pastern (just above the hoof). Therefore (when handling the legs and hooves, always warn the horse that you are about to ask it to lift a hoof by running your hand from the body down towards the leg.) Your knees should be bent slightly to reduce any strain on your back. Your back should be kept straight, to avoid strain. Simultaneously, you lean on the leg of the horse so that it shifts its weight to its other legs, and grasp its leg around the back of the pastern and lift the hoof, taking care to keep your thumb on the same side of the leg as your

a) Stand at the side of the horse and b) Run your hand from the hindquarter down the leg. c) Grasp around the pastern, lean on the horse and lift the leg

fingers. This prevents your thumb from getting injured if the horse lifts the leg too quickly.

As soon as the horse lifts the hoof it should be supported by your forearm across the back of the pastern and your hand under the front of the hoof. Your body should be kept close to the horse and your knees kept bent.

Once the hoof is securely supported, the hoof can be picked out with a hoof pick. Work the hoof pick from the back of the hoof, near the heels towards the toe area. This will dislodge any packed earth, manure and stones trapped in the hoof.

To pick up a hind leg, stand level with the flank of the horse facing towards the tail (having walked there from the shoulder, with the horse aware of your presence). Your nearest hand is placed flat on the hindquarter then run down the leg to the pastern. Again, always warn the horse by running your hand from the body down towards the leg. Your knees should be bent slightly as you bend over to reduce strain on your back. Grasp around the back of the pastern. Pull the leg forward to lift it off the ground. Again, take care to keep your thumb on the same side of the leg as your fingers while the hoof is lifted. You may find it more comfortable to rest the front of the hoof on your knee, otherwise, the hind hoof is supported the same way as a front hoof, with your body close to the horse and your knees bent. The hoof can then be picked out the same as a front hoof.

When the front hoof is released you should lower it gently to the ground. With the back leg, step backwards (towards the shoulder of the horse) and gently lower the hoof, in both cases making sure your feet are not where the horse will put the hoof down! Horses do not deliberately stand on feet; people put their feet in the way.

A horse may try to wrench its hoof away while you have hold of it. Ideally, you should hold the hoof until the horse relaxes and then release it. By doing this you are rewarding the horse for relaxing rather than rewarding it for wrenching its leg away. If you can't hold the hoof you must let go and step to a safe position next to the shoulder, if not already there. Then try again or get someone who is more experienced to do it until the horse is happy to have the hooves picked up and held up. If the horse is repeatedly allowed to snatch a hoof back it will become a habit, so get help early.

Be particularly careful when handling shod hooves that have risen clenches (the nails that hold the shoe on are called clenches; they 'rise' when the shoes are loose) because they can rip your hand if the horse snatches its hoof away.

Hoof care

Horses' hooves must be kept at the correct length and shape. Hooves that are too long can cause a horse to trip or stumble. Shod horses need to be reshod (either with new shoes or the old ones reapplied if not too worn) approximately every six weeks irrespective of whether the shoe is worn out, because the hoof continues to grow but is not worn down due to the shoe. When the shoes have been in place for approximately six weeks the 'clenches' (turned-over nails) start to rise. This means that the nails that come out of the hoof wall start to lift away and the shoe begins to loosen. Another sign that the horse is due to be reshod is that, when you look from underneath, the hoof begins to grow over the outside edge of the shoe.

Unshod horses may also need regular trimming depending on the surface on which they live. Soft surfaces such as grass and stable bedding do not wear the hoof down sufficiently and horses that live on them may need their hooves trimmed every six weeks. The gap between farrier visits for an unshod horse depends on hoof growth and the ground that the horse is living and working on.

Handling the mouth

There are times when it is necessary to open a horse's mouth such as when the teeth need to be inspected so that you can judge the horse's age however, there is no point in doing this unless you know what to look for. Before having its mouth opened a horse must be untied and the noseband strap of the halter must be loosened enough that the horse can open its mouth without the strap becoming tight. Horses have large gaps on either side between their incisors (front teeth) and molars (back teeth), called the bars of the mouth.

To open the mouth, slip three fingers into the area called the bars, moving across the tongue and aiming for the other side of the mouth. Then gently but firmly grasp hold of the tongue and pull it out of the mouth. The tongue will help to keep the mouth open.

The incisors can now be inspected, but your fingers must to be kept out of the way as the horse can still close the mouth although it will tend to keep it open. When you have finished, release the tongue and the horse will pull it back into the mouth.

This procedure is not something that an inexperienced horse person should do or even attempt. Besides which it takes a lot of experience to be able to judge the age of a horse by its teeth.

A horse's mouth of a horse also has to be handled when you are bridling it. Some horses will open the mouth without you having to do anything other than place the bit under the mouth. However, this is not common and you need to know how to open the mouth so that you can safely bridle any horse if necessary. In this case, open the mouth by placing the thumb of your left hand into the bar on the left-hand side of the horse while manoeuvring the bit carefully into the mouth with the flat of your left hand (see p. 79).

a) To open the mouth, put three or four fingers into the side of the mouth aiming for the other side. b) Take hold of the tongue and gently pull it out of the mouth.

Dental care

A horse needs regular dental care from a qualified equine dentist or vet

Teeth problems can cause behavioural problems as the horse attempts to alleviate pain. Domestic horses require regular dental care from a qualified equine dentist or veterinarian.

The teeth of a mature horse should be checked at least once every 12 months if it is predominantly grazing. Horses that are fed large amounts of concentrates (rather than having access to grazing) should have their teeth checked every six months. Young horses (between the ages of two and five) require six-monthly inspections to remove any caps (milk teeth that have not shed properly) and should always have their teeth checked before being mouthed (a bit in the mouth) for the first time.

Feeding

Horses require the correct level of feed to thrive. Underfeeding a horse is a welfare issue and leads to weakness, lethargy and ill health. Overfeeding can lead to a horse becoming difficult to manage, to the point of being dangerous. What to feed a horse is a complex subject that cannot be properly covered here. This section deals with the safety aspects of *providing* feed to horses.

Recommended further reading

Kohnke J (1998). *Feeding and nutrition of horses: the making of a champion*, 3rd edn. Vetsearch International, Australia.

Huntington P, Myers J & Owens L (2004). *Horse sense: the guide to horse care in Australia and New Zealand*, 2nd edn. Landlinks Press, Melbourne.

Providing feed to horses

Providing feed to horses is a potentially dangerous situation because horses naturally get excited when anticipating feed. If horses are in groups when you are handing out feed they will jostle and often kick and bite each other because giving feed to horses (as opposed to them foraging for their own) causes them to be more aggressive. By feeding groups of horses in confined spaces we create an unnatural situation – in nature, no one comes along with a bucket of feed. Of course, supplementary feeding is often necessary but it must be done safely both for humans and horses.

Whenever possible, avoid providing supplementary feed to groups of horses loose in a large area. It is impossible to feed different horses different amounts this way. Therefore, unless all the horses are the same class (e.g. all yearlings) and receiving the same feed they need to be brought in to receive individual feeds. It is far safer for all concerned to put horses into individual yards or stables for feeding. On a stud or other commercial operation this may not be possible because of time constraints. The next safest alternative for the people feeding out is to have feeders fixed to the fence so that the feed can be given out via the laneway without people having to enter the paddock with the feed. Feeders should be at

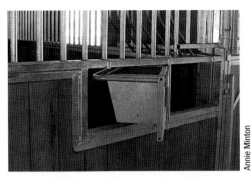

A swing-out feeder eliminates the need to enter the stable with feed

least 15 m apart along the fence line. The fence should be strong and safe; if electrified, that section should not have outriggers due to the risk of horses getting a shock while eating by simply raising their head. Any electric fencing in the area should be covered by a sleeve of poly pipe and fastened close to the fence.

Feeders in stables or yards should be positioned by the door or gate so that you don't have to walk out behind a horse that is eating. Swing-out feeders eliminates the need for someone to have to bring feed into the stable. If a feeder is in a fixed position at the back of the stable or yard, either take the horse out while the feed is put in the feeder or put a halter and lead on the horse and lead it to the feeder to put the feed in then lead it back to the door to release it (see p. 70).

Confined horses that get aggressive about feed are often being inadvertently rewarded for their behaviour. For example, if a horse is kicking the door or pulling faces as feed is prepared and brought, the arrival of feed is rewarding it for this behaviour.

If a horse behaves in this way at feed times, first make sure that the behaviour is not justifiable. For example, confined horses are often fed a diet that is far too low in fibre. This can result in a high level of unused stomach acids, which leads to gastric ulcers. Kicking and bad behaviour very often occurs because the horse is in pain and wants the food to come quicker to relieve the pain. Have a vet check the horse if it shows such behaviour. Ulcers can be detected and treated by a vet. Horses need a diet that is as close to natural as possible, namely a high forage/low- or no-grain diet.

All horses should be fed this type of diet so that the natural way of eating is mimicked as much as possible. A high forage/low- or no-grain diet ensures that the gut is never completely empty and the horse is not susceptible to problems such as gastric ulcers.

With horses that get aggressive about feed, change the routine so that feed is not prepared and brought to the horse, while in the stable or yard for example if the horse is turned out for the day have the food waiting in the yard or stable for its return. This will prevent the horse getting excited about the arrival of feed. If you can't avoid approaching the horse with feed, proceed only when the horse is still. If the horse kicks the door, stop walking, so that the horse will be retrained to not kick the door or walls.

Feeding by hand

Feeding horses by hand is a controversial subject among horse people. You should not give titbits to horses by hand unless you are experienced enough to know the right time to reward, i.e. during or within a second or two of a horse exhibiting a correct behaviour. Horses that are given titbits on an ad hoc basis learn to shove, nip or even bite. For example, if a horse nudges or grabs at your hand with its lips and then is given a titbit the horse will repeat this behaviour for more and more food rewards. In time the horse may start to nip or bite (common in small ponies due to being fed titbits at the wrong time) if the food is no longer forthcoming. Therefore, take care when feeding horses by hand as it will increase the chances of a behaviour being repeated, irrespective of whether it is good or bad behaviour. Food is a very strong reinforcer of both good and bad behaviours because it is high in the hierarchy of needs; a rub or a scratch is nice but is not the difference between life and death.

You should not enter an area that contains a group of loose horses to give them titbits because they can crowd around and fight each other to get the food – and you will be the innocent casualty. It is also not safe to feed loose horses over a fence as horses can fight each other and inadvertently injure you through the fence, for example a kick can reach over or through the fence. The horses are also likely to injure each other or get caught in the fence.

Avoid letting children feed horses by hand. This is difficult because children particularly like to do this. The dangers are that the horse can nip or bite the child's face or fingers, and that horses can learn to grab and snatch food when it is given to them in this way (young children will often pull their hand away due to nervousness as the horse reaches for the food, which leads to it 'grabbing' for the food). If children must be allowed to feed horses put the feed in a bucket and supervise while they hold the bucket. The horse should be on the other side of a solid barrier such as a stable door.

Never tease a horse, especially with food. When horses are eating leave them alone. Do not put on or take off rugs, for example, while a horse is eating a feed. Don't allow children or dogs to harass a horse while it's eating.

Rugging

Rugs are appropriate in some situations. They must fit well and be made of suitable materials, for example outdoor rugs should be waterproof. Poorly fitting rugs can restrict a horse's movements and therefore make it colder rather than warmer. The number of rugs should be varied depending on the weather conditions. Over-rugging can cause a horse to overheat in warm weather. This can become a serious welfare issue. Horses that wear rugs must be supervised and checked frequently because the rugs can slip and either injure the horse or prevent it from moving.

Fitting a rug

A rug must be put on and taken off with care. You must have someone else hold the horse, or tie it up, while you are putting on or taking off the horse's rug. Most modern rugs fasten at the front (across the chest) and at the back (via leg straps through the back legs). When putting a rug on or talking it off, there is always a time when a rug is only

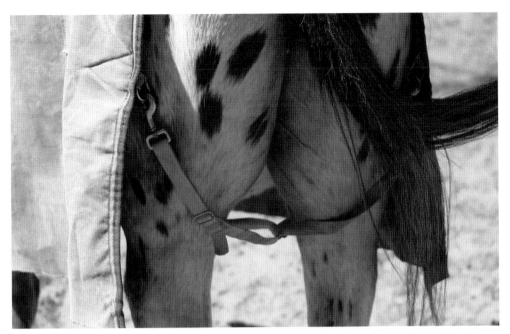

Leg straps cross through each other and fasten back to the same side. Clips should face inwards to reduce the chances of them clipping onto a wire fence if the horse leans or rubs on it.

partially fastened. This period can lead to accidents if the rug slips over or blows up before it is secured properly.

Some people feel it is safer to fasten the chest straps first when putting on the rug and unfasten them last when removing the rug, so that if the rug slips while the back leg straps are unfastened it is the back of the rug that will move. What tends to happen is that the rug slips around the horse's neck. Some horses panic at this and may pull back or run backwards. The other way of doing the straps is to fasten the leg straps first when putting on the rug and unfasten them last when removing the rug. This can result in the rug slipping around the back legs, which can cause the horse to panic and kick out. There are pros and cons to both ways of doing it but whichever way you choose always fasten or unfasten the other straps as fast as possible to shorten the time the horse spends with just the front or back straps fastened. If a horse does start to panic when a partly fastened rug slips, get clear of the horse if there is any chance that you could be knocked over or kicked.

To put a rug on, first check that the leg straps are fastened back to themselves so that they do not swing around and hit the legs or belly of the horse. The rug should then be gently placed across the back. When it is in place, fasten the straps immediately. Before you remove a rug, fasten the leg straps back to themselves *before* lifting the rug off the horse. If you don't do this the straps can catch on each other or in the horse's tail and the rug will be caught around the horse's back legs as you try to take it off. This is a common cause of horses kicking out as a rug is removed.

A horse with an unknown rugging history should be treated as an inexperienced horse before it is rugged for the first time in your care (see below).

One way to introduce an inexperienced horse to a rug is as follows. Bring the horse into a safe enclosed area. A well-built round yard with high fences is ideal. Never put a

The rug can be folded into a tube if you are placing it on a horse for the first time

rug on an inexperienced horse for the first in a large area such as a paddock. A horse can panic and run through a fence.

Before having a rug on for the first time, the horse should be habituated to the feel of items such as a saddle cloth and eventually a larger piece of material such as a tarp, and to the feeling of straps in various positions (see p. 128). If a horse won't accept something flapping around it or won't tolerate straps around the legs then it is not ready to wear a rug.

The horse should be held by someone who is capable (see p. 66). The rug should be folded into a tube with the front part on top. Check that the horse is not frightened of the sight and the feel of the rug by testing its reaction when you rub the rug against its side and back. If it is calm, slide the rug across the withers. Carefully unfold the front then fold the back part out and carefully fasten all the straps. Be very cautious when fastening the straps that go between the back legs and stand well forward, near the flank rather than further back. Leave the horse in the yard to get used to the feel of the rug. Make sure the yard has no projections that could catch on the rug.

After the horse has worn the rug for an hour or so catch the horse again and play around with the rug, slapping the rug and moving it about a little. This simulates the unfamiliar noises that the rug will make when the horse brushes against something. Do this until the horse is not worried by it, and turn the horse out into a pasture only when it has accepted the rug and is relaxed. Be aware that other horses in the paddock may treat this strange-looking 'newcomer' differently when it is reintroduced to the herd, even though they lived with the horse before it was rugged.

Parasite protection

The major parasitic 'worms' that affect horses are the large and small strongyles, roundworms, pinworms and bots. Horses should be wormed regularly (every 8–12 weeks depending on their infestation level and the property management practices) with an effective product.

Administering a worm paste

Administering a 'worm' paste can be a potentially dangerous situation if done incorrectly. Both the horse and the handler begin to see the process as unpleasant and react accordingly. The horse jumps around, lifts its head up and tries to avoid the applicator. It may even rear. The handler, anticipating this, can react to this behaviour badly and thus encourage the cycle of misbehaviour. Administering worming paste need not be such a harrowing or potentially dangerous experience if you do some preparatory work beforehand.

Administering a worm paste can be messy and an expensive exercise if the paste ends up on the ground, so it is a good idea to practise with an old paste syringe containing something that the horse likes, such as apple sauce, until you have perfected your technique. This has the added benefit that the horse begins to enjoy the process. To administer a paste have the horse untied but wearing a halter and leadrope. Make sure that the noseband part of the halter is not tight. Place the leadrope over the horse's neck and stand at the side of its head, with one hand on the leadrope. Wait until the horse has finished chewing any food and has swallowed (otherwise the paste sticks to the food and the horse can spit the whole lot out). Put the syringe onto the corner of the mouth, pointing towards the back of the mouth. Squeeze the syringe until it is empty.

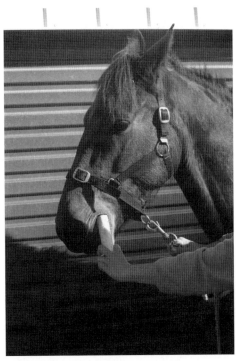

Administering a worm paste

Horses that are difficult to worm and have well-entrenched aversive behaviours may not even let you get near enough to administer something pleasant such as apple sauce. If you are inexperienced, get the help of an experienced horse person to retrain the horse to be calm when being wormed.

Drenching via a stomach tube is not needed these days as modern worm pastes that are administered by mouth have rendered this practice unnecessary. Some people still insist on having their horses drenched however, but this should only be carried out by a vet.

Inoculations and injections

There are various diseases that horses can be inoculated against but the most important is tetanus (lockjaw), which is usually fatal. Horses are highly susceptible to tetanus and the environment in which they live harbours the bacterium. This bacterium is anaerobic (it lives without oxygen) so certain types of wounds, such as deep punctures, are more prone to the development of tetanus.

Horses can also be inoculated against other diseases, such as strangles, but a commercial operator should consult with a vet about the economic viability of inoculating against strangles as the vaccination is not an absolute preventative. Also, many older horses have already been exposed to strangles and have developed some immunity. Horse owners should consult with their vet on the subject of inoculations as each person's situation is different.

Some inoculations can be purchased for you to administer. However, you should never inject a horse unless a vet or very experienced horse person has shown you how to do it. Innoculations are intra-muscular, which means that they go into the muscle. Intravenous (into the vein) injections should only ever be given by a vet.

Working alone

It is safer to have someone else around when handling horses, however, this is not always possible. People who are on their own should be even more observant and not take any unnecessary risks. A mobile phone should be nearby so that you can call for assistance if necessary. Higher-risk tasks such as administering a worming paste or injecting a horse should be left until someone else is present.

Assisting another person

Sometimes it is necessary for two people to handle a horse at the same time. Examples are when a farrier, vet or horse dentist is working on the horse and require the horse to be held. Some professionals prefer the horse to be tied up securely and left; farriers often prefer this. If you hold a horse for someone else you must understand that it is your responsibility to assist the other person and to help keep the situation as safe as possible.

The assistant must always stay on the same side of the horse as the other person and must not stand in front of the horse. This is because if the horse moves the assistant can then bend the neck towards them to help control the horse and the hindquarters will move away from rather than towards the other person. It is the assistant's responsibility to keep swapping sides as the other person moves around the horse. The assistant must keep alert at all times and not allow the horse to move its head excessively or to lower its head to graze or eat feed off the ground.

When two people are handling a horse they must both be on the same side

Restraining a horse

There are times when a horse must be restrained in addition to being held. This is usually because the horse is injured and needs a veterinary procedure such as an injection, sutures or wound dressing. As horses tend to panic if they feel trapped, restraint can make a situation worse if it is done badly. A well-trained quiet horse will respond better than an untrained or infrequently handled horse. The problem with restraining a horse is that horses naturally fight restraint because it prevents them from being able to carry out their primary flight response, i.e. to run away. In some cases it also prevents their secondary response, which is to defend themselves by kicking and biting. Even supposedly quiet horses can panic if they are inappropriately restrained. **Therefore restraint is not something that a novice horse handler should attempt.**

When a horse is injured its behaviour can change. Depending on the injury and the situation the horse may be quieter than normal or may be more anxious and difficult to handle. When a horse is stressed, try to have another horse nearby to help keep it calm. Keep the second horse in sight either in another yard or tied up nearby. Don't expect an already anxious horse to stand still if his companion walks out of sight.

A good strong yard or stable can be used to help restrain a horse if it is behaving reasonably quietly; again, this should not be attempted by a novice. The horse can be wedged with its back end in a corner. The person holding the horse stands facing the neck of the horse with one hand on the halter and the other pressing the horse's shoulders into the wall. The hand that is holding the horse's head turns the head slightly in from the wall. The other hand can be used to employ a tinker's grip (see p. 69). The person administering the treatment can then get on with what they need to do. At no time should either person let themselves get caught between the horse and a wall.

If the horse cannot be made to stand still safely using this position, it may be safer to hold the horse in a larger enclosed space (such as a round yard or square yard) so that people do not get trapped between the horse and a wall or fence. This way the horse can be allowed to move a little or even walk around if the pressure becomes too much. Sometimes just allowing the horse to move in a circle is enough to let it relax. Every situation is different, so the circumstances should be assessed by an experienced handler before deciding what to do.

Sedation is an option for immobilising a horse, but it should only be administered by a vet. Many vets understandably prefer an injured horse to be sedated before they work on it. Obviously, if the horse is behaving dangerously then it should be sedated. But it is not always necessary. A competent horse handler can help a vet by restraining the horse and keeping it calm. If the horse is to be sedated it may need to be restrained somehow so that it can be injected with the sedative. It is far easier to inject a sedative before the horse becomes stressed, both in terms of getting the sedative into the horse and the effectiveness of the sedative.

'Twitching' means applying pressure to various parts of the horse's body in order to restrain it. A common form of twitch is a nose twitch, which is pressure applied to the horse's top lip. The top lip has many nerve endings and is very sensitive. It is believed that when a twitch is applied the body releases natural painkillers that have a calming effect on the horse. When a twitch is properly applied the horse stands still with its eyes almost half-closed. At this stage the horse will accept an injection or many other procedures that may usually result in it being fearful (not all horses are frightened of injections etc.).

There are various ways of applying a nose twitch. The most common is a piece of poly pipe with a loop of soft twine threaded through a hole at one end. Avoid using a twitch with a wooden handle, because if the horse starts to swing its head around the wooden handle can hit you in the head. The twine is placed around the horse's top lip and is twisted until it is tight. Another form of twitch is made from smooth metal tubes; this type is very quick to put on and off. Some horses can be effectively twitched simply by a handler grasping the top lip very tight between the fingers and thumb.

It is safer to have two people when using a nose twitch, one to hold the horse and apply the twitch while the other one carries out the treatment. You can buy nose twitches

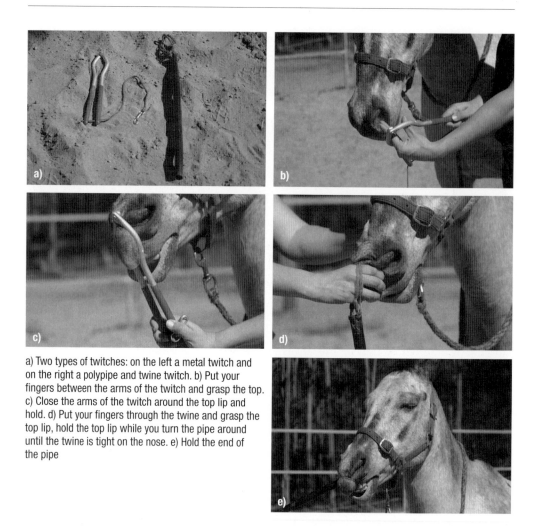

a) Two types of twitches: on the left a metal twitch and on the right a polypipe and twine twitch. b) Put your fingers between the arms of the twitch and grasp the top. c) Close the arms of the twitch around the top lip and hold. d) Put your fingers through the twine and grasp the top lip, hold the top lip while you turn the pipe around until the twine is tight on the nose. e) Hold the end of the pipe

with a clip so that they can be attached to a halter, enabling one person to work alone with the horse. However, this is not as safe.

Opinion is divided about whether twitching is a cruel practice. Correctly used, a twitch can save the horse a lot of stress. A good example is when the horse needs frequent cleaning of a wound, in particular, taking care of an injured eye. It is not possible or desirable to tranquillise the horse several times a day for such a procedure. Using a nose twitch, if necessary, lets you dress the injury with minimal stress to the horse and increased safety for the handlers.

A twitch should not be used simply to make the horse stand still if it has not been trained properly. Twitches are often used for routine tasks such as trimming, clipping and shoeing, when they should not be necessary. Understandably, some horses do not like being clipped around the head because the head is largely hollow and the sound of the clippers becomes very loud indeed. Instead of using a twitch, the horse should be habituated (see p. 128) to the sight, sound and feel of the clippers.

A twitch will only work for a certain length of time, which varies from horse to horse. When a horse is excited or frightened it releases adrenaline into the bloodstream. This hormone can override the effect of the twitch, either making it ineffective or even making the horse react to the twitch. For this reason it is safer to apply the twitch before the horse gets upset, quickly administer the necessary treatment and remove the twitch. A twitch should always be used for the shortest possible time. In skilful hands, administering an injection may take only one or two minutes if everything is prepared beforehand as it should be.

Another form of twitch, which is milder but can be very effective for some horses, is where the skin on the neck (where the neck joins the shoulder) is grasped, twisted and squeezed. This has a similar but weaker effect than a nose twitch. This form of twitch has various names worldwide. One name for it is a tinker's grip.

Holding an ear can help to restrain a horse, however a twitch (such as that used for the nose) should never be applied to the ear. Horses can become ear-shy very easily if their ears are mishandled so take great care with them.

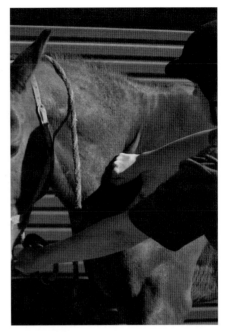

A tinker's grip

Covering the eye with your hand can be effective to prevent a horse from seeing behind. This can be useful in the case of injections, especially if the person administering the injection is quick and skilful.

There are various methods of leg restraints used by horse trainers, with differing levels of success. They include knee hobbles, tying up a front leg, a sideline etc. Leg

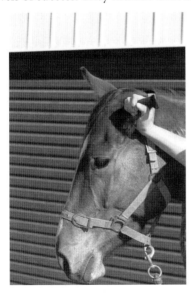

Holding an ear

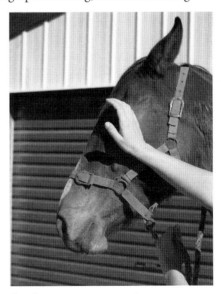

Covering an eye

restraints should only be used to restrain a horse if it has been habituated to their use previously and in these cases they can be invaluable. They should *not* be applied to a horse that is relatively unhandled and when it is in a stressed state of mind. In that case the horse will panic further and can become dangerous.

A useful way to think of leg restraints is that all horses should be habituated to their use *after* having been handled/started and when they are already quiet and unlikely to panic. A horse that is habituated to leg restraints is less likely to panic in situations such as being caught in a fence or a rug slipping around its legs. It also means that, if necessary, the horse can be restrained in an emergency situation such as when injured. The fact that the horse has learned to yield the legs rather than fight the pressure can be a life-saver for the horse and dramatically increase the safety of its handlers.

Releasing a horse

There are several potential dangers involved with releasing a horse.

- The horse gets excited as it anticipates being released into a grassy pasture or a yard/stable that has feed in it.
- The horse can kick out as it passes you after release.
- You can be injured when releasing a horse into an area that already contains horses if they crowd the gateway.

Safe procedures reduce the chance of accidents.

When you lead the horse into the area that it will be released into, turn it back to face the door or gate. Gates into pastures should swing either way or outwards only (see p. 33).

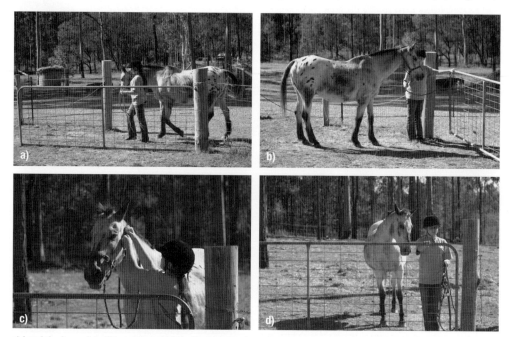

a) Lead the horse into the paddock. b) Turn the horse to face the gate and pull/push the gate to close but don't fasten it. c) Take off the halter. d) Walk straight back out of the paddock

The door or gate should not be fastened because you need to be able to get out quickly – the horse should be released and then you step out of the area immediately. A horse can kick out in excitement when first released, so by turning the horse you have a slight advantage. When releasing a horse into a stable or small yard, turn the horse back to the door before releasing it. You should always walk away from a horse first, not the other way around.

You should never slap a horse on the rump after letting it go! As well as being very dangerous, you will be teaching the horse to run away. **Good training is about suppressing the flight response, not encouraging it.**

If several people are going to release several horses simultaneously into an area, you can use the following method. Lead all the horses about 20 m into the area after closing and fastening the gate. Turn all the horses to face the gate in a large fan. All the horse handlers then undo the halters, leaving the rope high up around the neck of each horse. The handlers then hold the horses with this rope. A previously appointed leader gives the command to release the horses at the same time, after checking that everyone is ready. As soon as the horses are quietly released the horse handlers should walk straight out of the area.

Common handling problems

Horses that do things such things as knock handlers over, barge past them when handled on the ground and so on behave that way because they have not been trained to move away from pressure when handled (see p. 125). Once a horse is trained to move forwards, backwards and sideways on the ground the handler has the tools to make sure that the horse does not invade their space, walks quietly beside them etc.

Other handling problems include aggressive behaviour directed at humans or other horses. Domestic horses can become overly aggressive for many reasons.

- Instead of feed being everywhere, as is grass, domestic horses are often fed 'meals', which can result in horses fighting over food.
- Confining a horse in a small area means that the horse is cornered rather than free to move away from a perceived threat; a frightened horse has to defend itself with aggression rather than use the preferred option, which is to flee.
- Inappropriate cues (i.e. clumsy/inconsistent/unrealistic instructions) lead to frustration and confusion in the horse, and the horse may become defensive or aggressive.
- Some horses may learn to become overconfident and aggressive if they are allowed to move a human around as they would another horse.

Nipping and biting

Many bites are actually the result of inexperienced people getting their fingers in the way rather than the horse deliberately biting them. A common scenario is people feeding horses by hand and not keeping their fingers away from the teeth. Feeding titbits by hand can be the start of the horse learning nipping and biting behaviour. As a general rule, novice horse people should not feed horses by hand even though experienced horse trainers may choose to use food rewards (given by hand) as a training tool (see p. 62).

Aggressive biters should not be handled by inexperienced handlers – this is very dangerous behaviour. As with all problem behaviours, the horse needs to be retrained by a professional so that it stops biting and even then only experienced horse handlers may be able to handle it. The legal responsibility for a known biter lies with the owner. Even if the owner isn't there or if an intruder gets bitten, the owner is fully responsible.

A horse bite can do a lot of damage. The horse bites with its incisor (front) teeth and, because these teeth meet evenly at the front of the mouth and the horse has very powerful jaws (for chewing grass for many hours a day), the injury is more akin to a crushing injury than a puncture. In contrast, a dog uses sharp canine teeth to puncture and rip the skin. A horse bite can also puncture the skin, of course, but the worst damage is from crushing.

A well-trained horse generally won't bite a human because it has been trained not to and usually has no reason to, however, handlers must always be on their guard especially when handling unfamiliar horses.

Horses nip or bite for many reasons, including those listed here.

- Age and sex – young horses, especially colts, are 'mouthy'. Young horses tend to taste new things in their environment. This can include anything in their vicinity, including people.
- Incorrect handling and training.
- Anxiety – a horse that feels cornered may bite in defence.
- Irritation, such as at a fly or when being girthed up.

Young horses are naturally mouthy and will nip and bite other horses when playing. This is normal play/learning behaviour for a young horse and other horses will reprimand it (with a bite or a kick) if it goes too far. Other horses in fact teach the youngster good manners. When young horses are kept alone they are much more mouthy than when allowed to play with other horses. This is one of the reasons why young horses especially should not live in isolation. They need to interact with other horses to learn socialisation skills.

A young horse should not be allowed to play with humans in the same way. Any nipping or biting should be firmly discouraged by pushing the nose away. Always have at least a halter and leadrope on a horse if in biting range. Never play with the mouth of a young horse other than to accustom it to being handled in this area, because this encourages mouthy behaviour. Slapping a mouthy colt can lead to a game (to the horse) that involves the horse swinging the head away from the slap then back to bite. Young colts play a similar game by biting at each other and if you're not careful it is possible to set up a situation where the horse thinks you're playing the game. A handler should either stand out of reach or the horse should have a halter and leadrope on so that you can control the head.

Young horses need very careful handling if they are not to learn bad behaviour. They should not be handled by inexperienced people, who may teach them bad habits that will be difficult to eradicate.

A horse that is anxious may bite even if it has not been known to bite before. Biting is a natural method of defence for a horse. If the horse is pushed into a corner either literally or otherwise it may feel that it has to defend itself because it cannot run away. For example, a horse that is anxious on a horse trailer may nip or bite anything within reach in that situation. A mare will bite to defend her foal if she feels it necessary.

When a horse is being groomed or scratched, it may try to 'groom' the handler with its incisors. Horses groom each other in this way (reciprocal grooming) and the horse is not being nasty. If this happens, just push the head away as you groom or scratch and the horse will get the message.

Two horses grooming each other

Many horses nip or bite when being saddled. They form this habit when first learning to wear a saddle because the saddle was girthed up too tight and too fast. The horse responds to this by nipping or biting, which becomes an entrenched habit (see p. 92).

When handling a horse pay attention to the body language as the horse usually warns before biting – it will lay back its ears and swing the head around in the direction of the handler. Always wear loose clothing when handling horses that nip or bite, as this reduces the chances of the horse's teeth catching you.

Kicking and striking

Many kicks are the result of people startling the horse by approaching from the wrong angle. Horses that are known to kick should only be handled by experienced horse people. The legal responsibility for a known kicker lies with the owner.

As with nipping and biting, horses kick for a variety of reasons. They all have the potential to injure just as much. Typical kicking scenarios include those given here.

- Kicking from fear because the horse cannot escape a frightening situation. An excessively nervous horse or a horse that has little previous handling will kick from fear if it is cornered. In this case the horse has no choice but to defend itself. The handler must get out of the way. This kind of horse should only be handled by an experienced horse person who knows when and how to apply pressure and when to remove pressure safely.
- Kicking from overconfidence because the horse has learned that it can get away with it. This kind of kick ranges from a half-hearted kick out as the horse passes a handler, to a full-blown double-barrel kick (both back feet). This kind of horse wasn't handled properly during its basic training and it has learned that it can kick without being reprimanded. This is often the result of being kept on its own. Horses that live in a herd learn socialisation skills which include curbing kicking behaviour. Isolated horses, unless handled expertly, miss out on etiquette lessons.
- Kicking at another horse and a human gets in the way. This is a common scenario which can be prevented. Common situations include one horse kicks at another while being ridden, or you are leading one horse out of a paddock and another horse kicks yours. Sometimes when horses are greeting each other, such as through

the bars of a stable wall, they will strike forward with a front leg but sometimes they may kick out with a back leg.

- Kicking from sheer exuberance. An otherwise well-mannered horse can kick due to excitement, such as when it is released after being confined for a time. Another common scenario is when a horse is being worked on the lunge or loose in a round yard and gets too close to the handler.
- Kicking at flies on the belly, the girth or an itch. Some horses are excessively ticklish or excessively irritable. This kind of kick is aimed forward at the belly. Although not as deadly as a backwards kick (unless the handler has the head low down) it can still cause injury.

If a horse is known to kick other horses it should not be taken out in company if horses are likely to come up behind. This horse should always be ridden at the back of a group ride. It is not good enough to stick a red ribbon on its tail and tell everyone that it kicks. Unfortunately, some horses become kickers in company because other horses have been allowed to come too close and have stepped on the heels of the horse in front. This can result in the injured horse becoming a kicker. To prevent this, always keep a horse's distance away and make sure that no one gets too close to the back of the preceding horse in a group (see p. 114).

Any horse can kick if provoked so you should always be on your guard, but take even more care with a known kicker. To avoid being kicked by any horse, handlers should move carefully around a horse (see p. 53) and watch the horse for signs before kicking (see p. 10). Never go behind a known kicker and watch that other people don't either. Never let children or inexperienced handlers handle a known kicker. Keep it away from situations where it is likely to harm someone or another horse. A horse that kicks is a very dangerous animal and must be trained to not kick. This may mean taking the horse to a professional.

A horse can also strike forward with a front leg. Horses often do this when greeting another horse. Never stand in front of a horse and do not allow a horse to 'meet and greet' with another horse while you are holding it as, in addition to the striking danger, your horse's attention will be diverted to the other horse rather than being on you.

Never pass to the other side of a tied horse by ducking under the neck because the horse can strike forward with a front leg (see p. 53).

6

Safe gear and tacking up

This chapter describes the selection of safe gear (tack) and how to fit the gear to the horse for safety. It also describes the selection and fitting of gear, such as a saddle, for rider comfort and safety.

A horse cannot work well if the gear does not fit properly. Gear that rubs causes sore spots, leading to an uncomfortable horse. An uncomfortable horse is potentially an unsafe horse. Poorly fitting gear can be the start of many problems as the horse may not work well if it is uncomfortable or in pain. Fitting gear, particularly saddles, can be difficult until you are experienced enough to know what to look for. Some points are given in the following sections, but if you are inexperienced in this area also get advice on fit from people such as staff at the saddlery store where you bought the gear.

There is a huge variation in quality and price of gear. To the uninitiated this is confusing because cheaper gear may look just as good as more expensive gear. There are, however, some important differences that you should take into consideration. The reason that some gear costs less to buy is because it costs less to make. Inferior materials may be used, which results in an inferior product. In these situations it is not unusual for leather parts to rip or snap, stitches to rot or come undone or metal parts to break under pressure. Another problem with cheap gear is that it often does not fit well, resulting in a sore horse and rider.

It is possible to buy good gear second-hand, especially saddles. If you are on a limited budget it is usually far better to buy a good-quality second-hand saddle rather than spend a similar amount on a new but inferior saddle. A good-quality saddle lasts for generations. So being second-hand is no problem as long as the saddle has been looked after well and is not damaged.

When buying gear you can choose between synthetic or leather. Synthetic gear has gone from strength to strength since it was first developed. It has many advantages over leather in that it is stronger, it does not require the same level of care, and it can be used in wet weather without spoiling. It usually costs less than leather gear of similar quality because the synthetic material is less expensive to produce. The down side of synthetic gear is that it does not 'breathe' as well so where it is in direct contact with the horse it can cause heavy sweating.

The stirrup leathers (straps), reins and girth are all items that take a lot of strain and are vitally important pieces of equipment. Never economise on these items. Fortunately it is possible to buy all these items in synthetic materials, making them much cheaper and just as safe if not safer than their leather counterparts. If you cannot afford good-quality leather items then synthetic items are better than poor-quality leather.

Bridles

The primary purpose of a bridle is to control the head of the horse during riding, but it can also be used for leading a horse. Most bridles are used in conjunction with a bit but some are not (hackamores and bitless bridles). Like most horse gear they are available in synthetic materials as well as leather. There are some differences between English, western and Australian styles.

Irrespective of the style a bridle must be strong enough to withstand wear and tear. It must not cause discomfort to the horse and it should be easy to put on and take off.

Bits

Poorly fitting and inappropriate bits cause many problems. There are many different types, some of which are very severe. Even a mild bit can cause pain and discomfort in the wrong hands; a severe bit can be dangerous as a horse may react adversely if the bit is used inappropriately. **Instructors and parents should note that a horse that needs a severe bit is not a beginner's horse.**

Two stainless steel jointed snaffle bits. One has smaller rings and a thinner mouthpiece. The other has larger rings and a thicker mouthpiece

The most common bit is the snaffle bit and even this has many variations ranging from mild to severe. They might have no joint in the mouthpiece, or have a jointed mouthpiece. Curb bits are bits that have leverage and they are sometimes used in conjunction with a chain that goes behind the horse's jaw (a curb chain). The leverage is achieved by the reins being fastened at the ends of shanks so that when they are pulled the bit twists in the mouth, putting pressure on the horse's poll. If a chain is fitted, pressure is also put on the back of the horse's jaw. Curbs are usually unjointed in the mouthpiece, however some curb bits have a jointed mouthpiece.

Generally, an unjointed snaffle bit is the least severe, followed by the most commonly used bit which is a jointed snaffle. A curb bit is more severe, especially if it has a chain attached. A jointed curb is the most severe of all. Better training, not a more severe bit, is usually the solution to a problem such as not stopping or turning properly. You should never use a different type of bit if you do not understand how it works.

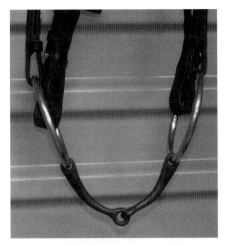

A jointed snaffle that shows the correct curve when fastened to a bridle. If it was fastened the other way it would nip the horse's tongue

The bit should have no sharp edges and should fit well. Bits can be made from various materials. Avoid using cheap metal-plated bits as the metal will flake off over time, and as they wear they develop sharp edges and can even snap.

If a bit is jointed make sure that it does not nip when the joint closes. Bits that have curved bars are much more comfortable on the tongue than those with straight. Very thick bits are not necessarily more comfortable than thin bits because they don't allow the horse to close the mouth properly, especially if a noseband is fitted that prevents the mouth from opening.

When in the mouth, the bit should not pinch the side of the mouth nor should it be so wide that it slides from side to side when the reins are used.

An area of contention among many horse people is how high the bit should sit in the mouth. There is no hard and fast rule because even horses have different preferences. However, the bit should never be low and loose in the mouth. Sometimes when a bit is too low the horse will actually hold it up in a more comfortable position, giving the impression that it is tight enough. Get the horse to open its mouth; if the bit drops down it is too slack. If the bit is too tight it will pull the corners of the mouth up. In this case the cheekpieces will also feel tight against the cheeks. You should be able to lift them 2 cm away from the sides of the face.

Reins

The reins should be comfortable in the rider's hands, therefore different riders prefer different thicknesses and materials. Reins are available in several materials, such as plain or plaited, leather, rubber-covered leather, cotton web or rope and nylon webbing.

Reins should not be long enough for the buckle end to catch around the rider's foot. For you to be comfortable, the buckle should sit about halfway between your hands and your foot when you hold the reins with a reasonable contact. If the reins need to be tied up in a knot at the end they are too long and should be replaced or altered. Open-ended or split reins (western reins) should be long enough that they don't slip off the neck if dropped across the neck. **Instructors and parents should note that beginner riders are usually safer with joined reins**.

Nosebands

A bridle may or may not have a noseband fitted. Apart from cosmetic reasons, a noseband may be fitted to prevent the horse from opening its mouth and evading the bit or it may serve as an attachment for a standing martingale or tie-down. In the last two cases you must get expert advice on fitting such gear, otherwise the horse may react adversely.

Correct length joined reins.

Correct length split reins

The simple cavesson noseband should be adjusted so that it sits about two-fingers width below the projecting cheek bone. It is fitted *under* the cheekpieces not over them. There should also be room to fit two fingers between the noseband and head. Be especially careful with a young horse (up to five) that may have molars erupting. Some riders fit this type of noseband very tightly so that the horse cannot open the mouth and evade the bit. In this case the horse should be educated to accept the bit better instead of having to rely on a very tight noseband.

There are various types of nosebands specially designed to keep the horse's mouth closed, such as a Hanoverian noseband. This noseband is a combination of a cavesson noseband and an extra strap. A horse that opens its mouth is not accepting the bit properly. Some of the possible reasons are a lack of training; poor riding technique, an inappropriate bit or even pain due to dental problems (see p. 120). All these issues should be addressed and if necessary rectified before you resort to putting a tight noseband on a horse. Such nosebands should only ever be used as a transitional training tool and not a permanent fixture.

To fit a Hanoverian noseband, the cavesson part of the noseband (the top strap) must be

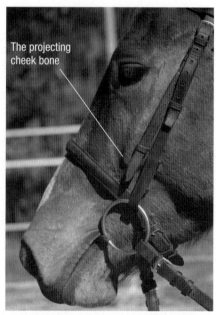

The projecting cheek bone

A correctly fitted cavesson noseband that sits two-fingers width below the projecting cheek bone

done up first, snug to the face (rather than leaving a gap for two fingers as for when it's on its own) then the other strap should go around the outside of the bit rings, again fitted snug to the face. This lower strap should not pull the cavesson part of the noseband down onto the bridge of the horse's nose.

A correctly fitted Hanoverian noseband. The lower strap is not pulling the top strap down the front of the nose.

Instructors and parents should note that a horse that requires gear to prevent it from opening its mouth is not a beginner's horse.

Browbands

The browband can be a source of discomfort to a horse if not fitted properly. If it is too tight it can pull the headpiece too close to the ears. Horses that are uncomfortable due to

A correctly fitted browband

an ill-fitting browband may shake their head, although some show no outward signs even though it is too tight. Some western bridles omit this piece of equipment, but they usually have a slot in the headpiece that fits over the horse's ear, to prevent the bridle coming off.

Bridling

Before putting a bridle on a horse make sure the head area is clean. Putting a bridle on a horse takes some practice, so learn on a horse that is experienced and quiet. If you are clumsy the horse won't like it and will eventually react against being bridled. There is more than one way of putting on a bridle; this is just one of them. The main safety considerations are that you should never stand directly in front of the horse when putting on a bridle (to reduce the chances of the horse hitting you with its head) and that you should open the mouth safely for the horse to accept the bit (avoiding bitten fingers).

- Check that the bridle has been put together correctly, and that it is the correct bridle for that horse.
- Carry the bridle to the horse on your shoulder.
- Standing on the horse's left side, untie the horse and place the rope over its neck.

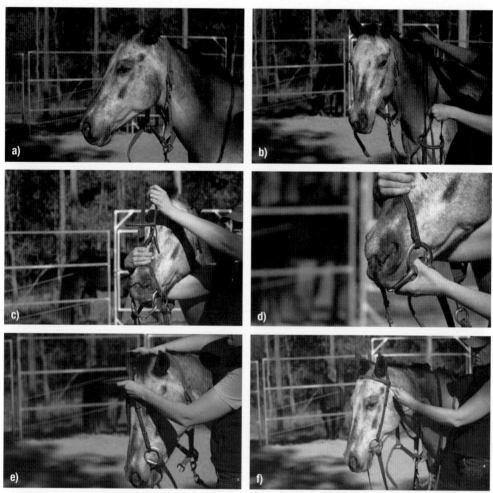

a) Unfasten the noseband of the halter b) Take the bridle off your shoulder and put the reins over the horse's head. c) Lift the bridle up along each side of the horse's face, taking up the slack with your right hand so the bit is lined up with the horse's muzzle. d) Slip your left thumb into the left side of the mouth and press on the bars of the mouth. Put the bit in the mouth, being careful not to bang the teeth. e) Lift the headpiece and put one ear then the other under it, preferably the offside ear first. Tidy the mane and forelock by grasping them firmly and pulling them out from under the straps. Make sure the bit, browband and noseband (if there is one) are level. f) Fasten the throat lash at a suitable tightness. If the bridle has a cavesson noseband, fasten it under the cheekpieces so that two fingers can fit between the noseband and the jaw.

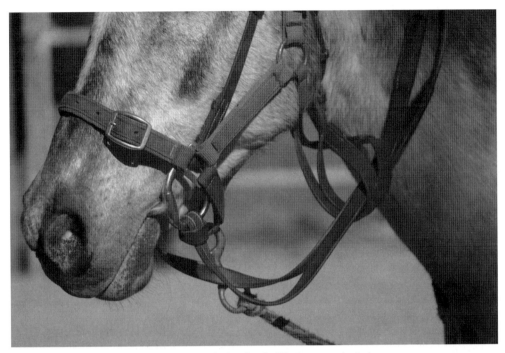

Secure the reins and place the halter back over the head so that the horse can be tied up

Once bridled, never tie the horse up by the reins. A horse can pull back and break its jaw. Also the horse is not as secure because bridle reins are not as strong as a leadrope. To tie up a bridled horse, take the reins back over the head and secure them by passing them around the neck, over themselves for one or two turns and then put the throat lash through the buckle end. Then put the halter back over the bridle and re-tie the horse. If the horse is to be re-tied frequently, it is quite acceptable to keep the halter on under the bridle the whole time.

Removing a bridle

If the horse is tied up (with a halter over or under the bridle and a leadrope) untie the horse before removing the bridle. A horse should not be tied up when a bridle is being removed because it can pull back as the bridle is taken off.

To remove a bridle, fasten a halter around the upper neck with the nose strap left undone. Place the leadrope across the horse's neck ready for you to get hold of after the bridle has been removed (see photos on p. 82).

Problems with bridling

If a horse is difficult to bridle, check whether its ears are getting pinched when the bridle is put on. Often a horse will start to resist if this has happened a few times. Let the bridle down a couple of holes so that it goes over easily, then readjust. Sometimes the browband is too tight (short) and all that is needed is a longer browband.

a) Undo the throat lash (and noseband if there is one). b) Take the reins up to the headpiece and lift the bridle carefully over the horse's ears. c) Wait until the horse opens its mouth to spit out the bit. Do not pull the bit out of the horse's mouth – this will quickly lead to the horse starting to panic when you lift the bridle over the ears as it will anticipate the bit banging its teeth. Young children and beginner riders should be supervised when removing a bridle as they may pull the bridle off too quickly, banging the horse's teeth causing it to panic. d) As soon as the bridle is removed put it on your shoulder and fasten the noseband of the halter around the horse's face. Tie the horse up and put the gear away in the tack room

Some horses will not open their mouth for the bit. This is usually because someone has previously banged the teeth when putting the bit in. This then becomes a vicious circle as the horse clenches its mouth and makes it more likely that the teeth will be banged again. If the horse will not open its mouth when you press your thumb in the gap behind the incisor teeth, wrap a piece of bread around the bit (squash it firmly with your hand). This has the twofold effect of eliminating the clinking noise that the horse associates with the bit touching the teeth, and making the horse want to take the bit because of the bread.

A horse that lifts its head up high to avoid the bridle should be trained to lower the head by pressure on the halter (see p. 125) or poll. Initially this should be done without trying to put the bridle on. Once this new behaviour is well established the bridle can be reintroduced. The horse will now be much easier to bridle.

Martingales and breastplates

Martingales help to prevent a horse from getting its head above a certain level, however they should be fitted and used by experienced horse people only and should be regarded as a stopgap until the horse is trained to travel with its head in a lower position. **Instructors and parents should note that a horse that travels with its head too high is not suitable for a beginner.**

A correctly fitted running martingale

There are two types of martingale, 'running' and 'standing'. The running martingale has rings that can slide along the reins and the standing attaches to a cavesson noseband under the horse's jaw.

A running martingale should be fitted so that the rings do not cause a dip in the reins when making a straight line from the horse's mouth to the rider's hands. The rings should only come into play when the horse lifts its head too high. Rubber stoppers **must** be used on the reins just after the billets or buckles so that the rings don't get caught in them. If that happens, the horse may panic at the sudden downward pull on its mouth.

Standing martingales attach to a cavesson noseband. Again, they should only come into play when the horse lifts its head too high. Standing martingales are more severe than running martingales because a horse can panic at the unyielding downward pull if it is not introduced very carefully. If you compete, check the rules as they vary widely on martingale use.

A breastplate will help to keep the saddle from slipping back in situations such as riding in steep country, jumping, racing or any other event that warrants it. A martingale can be attached to a breastplate if the breastplate is designed for this.

There are many different types of breastplate and they must be fitted so that they do not restrict movement of the neck or shoulders or interfere with the windpipe. The breastplate can be covered in sheepskin in the neck area for extra comfort.

Stoppers **must** be fitted to the reins if you use a running martingale to prevent the rings from catching on the billets or buckles.

Saddles

There are many different types of saddle to suit the many different disciplines of horse riding. Some of the common types of saddles are described here.

- English saddles (European) are used for the Olympic disciplines but are also widely used for everyday riding. This group includes dressage saddles, jumping saddles and all-purpose saddles which are a combination of jumping and dressage saddles. All these types of saddle have padded panels that should conform to the shape of the horse's back.
- Western saddles also vary depending on the discipline they are designed for. These saddles usually have a horn at the front for roping cattle, because this was their original purpose. They have flat rather than padded panels and therefore must be used with thick saddle blankets for padding.
- Australian stock saddles evolved from English saddles. Again, there are variations within the type. Like the western saddle, they are designed for comfort and rider stability when riding in steep country or when sudden changes in directions

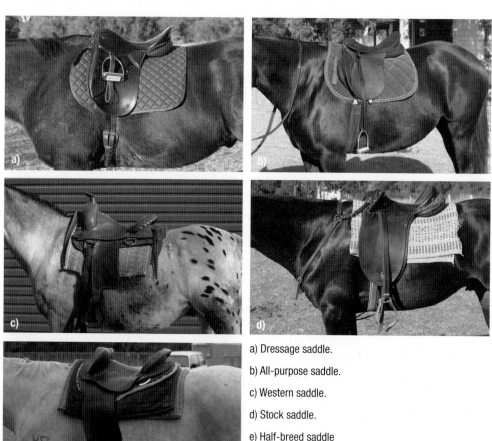

a) Dressage saddle.

b) All-purpose saddle.

c) Western saddle.

d) Stock saddle.

e) Half-breed saddle

(chasing cattle) are the norm. They are often used in conjunction with a breastplate. They have padded panels like an English saddle.

- Half-breeds are a combination of western saddles and Australian stock saddles. These are becoming increasingly popular with pleasure riders because they have the comfort of a western saddle but are hornless like a stock saddle.

The type of saddle you choose depends on the type of riding you are planning to do. It is important that the type of saddle used is suitable for the activity. Therefore a western saddle, Australian stock saddle or dressage saddle is not suitable for jumping, for example. Some people have more than one type of saddle, for their different disciplines. No one type of saddle is safer than the others. They have all been developed for specific styles of riding, however, some saddles are more versatile and some are not versatile at all, such as a racing saddle.

Stirrups

Stirrups help you stay on the horse. They are made from various materials such as stainless steel, strong plastic, wood, metal over wood, plastic, leather over wood or leather over plastic. Avoid cheap metal-plated stirrups which can snap or squash your foot in an accident. There is a huge variation in styles. Some are made to suit English saddles and some are made to suit western or Australian stock saddles. Some are traditional stirrups and some are classed as safety stirrups. The main danger with stirrups is that your foot can go though them and you can be dragged if you fall off. Some safety stirrups are not as safe as they seem, for instance, some collapse when you don't need them to, and others are more hazardous for entrapment than ordinary stirrups. Obtain advice from experts before spending lots of money on special 'safety' irons. Stirrups that entirely prevent your foot from going through are probably the safest. Some types of stirrup have a 'cage' around the front to prevent your foot from going through. Plain irons can be fitted with safety devices such as clogs or Toe Stoppers™ to prevent your foot from going through. There is also a type of hooded oxbow stirrup (called western tapaderos) that prevents the foot from slipping through.

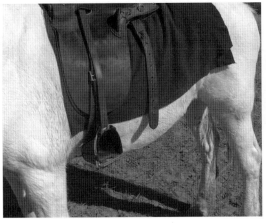

AHSE

Irons with Toestoppers™

Horse Connection

Western tapaderos

a) A wood (leather-covered) stirrup on a western saddle. b) A metal stirrup on a half-breed saddle. c) A caged stirrup. d) A stainless steel safety iron – the bend decreases the chance of your foot getting caught

A correctly fitted iron should have 1 cm on either side of your foot when the ball of your foot is on the tread. If the iron is larger than this your foot can slip through the iron too easily, if it is smaller your foot can be trapped. Both situations are very dangerous, if not deadly. Attachments such as Toe Stoppers™ make a large stirrup safe for a small boot as the boot cannot get caught in them, but there is a limit to this. The stirrup should still not be overly large.

Steel stirrups should be fitted with rubber treads (if not using clogs or Toe Stoppers™) to make them more stable under your foot. Rubber treads used to stop your foot slipping from the stirrup should not reduce your foot room.

Rubber tread on a stainless steel iron

a) Run the stirrup up the back strap.

b) Pass the leathers through the iron.

c) The stirrup will now stay in place

Being dragged is an accident that is highly preventable if you wear correct footwear (see p. 21) and use correctly sized stirrups. Using safety aids such as covered stirrups can reduce the chances even further.

Stirrups should be 'run up' whenever they are not in use. If this is not possible, such as with a western saddle, the stirrups should be put over the seat when not in use. This is to prevent the stirrups from swinging and hitting the horse, catching on projections or catching a horse's back foot if it kicks forward at a fly.

Stirrup leathers/fenders

Stirrup leathers or straps are so named because formerly that is what they were made from, however they are now made from either leather or synthetic material. Synthetic leathers are stronger and do not stretch like leather, but good-quality leather does not stretch as much as poor-quality leather. You can buy a combined leather and synthetic strap, which gives the nice look of leather with the strength of synthetic material.

On a western saddle the stirrup leathers are called fenders and are usually made of leather. They are much wider than stirrup leathers and have a different fastening system which is situated neat the stirrup rather than higher up under the rider's thigh. This is a two-part metal construction called Blevin slides.

Older English and Australian stock saddles have stirrup bars that can be either up or down. It is now regarded as safer for the bars to be down at all times to increase the chance that the leather will come off the stirrup bar if a rider is dragged. A consideration for stock saddles is that when the surcingle is in place, the stirrup leather may be prevented from slipping off the stirrup bar in the event of a rider being dragged. If this is at all likely, closed front stirrups should be used.

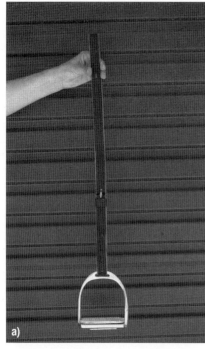

a) One type of synthetic leathers, called webbers. They have a slot and t-shaped metal fastener rather than a buckle. b) To run webbers up, take the stirrup *above* the highest slot and fasten it. This prevents the stirrup from coming down

A modern type of stirrup bar

An older type of stirrup bar in the down position

Girths/cinches

A saddle should be secured to a horse via more than one point of attachment. English saddles must have two girth straps, an Australian stock saddle must have a girth and a surcingle and a western saddle must have a double-wrapped latigo. There are numerous types of girths (English) and cinches (western) for horses – the main considerations are that they are the correct length and don't rub, as that will cause girth galls (pressure sores). Traditionally girths were made from leather, string or canvas. Leather was then the best choice but because it was also the most expensive, string or canvas girths were often

A stock saddle has one girth strap and should also be fitted with a surcingle, giving it two points of attachment

A saddle with long girth straps and a short girth. The horse's ribs are protected from the buckles with this particular type of girth

used instead. Cheap string girths (which were actually made from nylon) tended to cause galls, as did canvas girths unless kept very clean. Modern girths are made from various materials including synthetics. String girths are now better-quality girths that use wider and softer string or rope. Many girths, including leather girths, now incorporate elastic, which is far more comfortable for the horse and means that the girth does not have to be ridiculously tight to keep the saddle on. This is especially useful on horses and ponies that are round in shape as the saddle tends to slip more easily on those animals.

Some English saddles have long girth straps and some short. Long girth straps (points) mean that the girth buckles are not up under the rider's leg. In this case the girth should always have protection between the buckles and the horse's ribs.

The girth should ideally fit the horse when it is halfway up the girth straps; this way there is room for expansion or reduction over time.

Saddle cloths and blankets

A saddle cloth or blanket can be used to keep the underside of the saddle clean and to soak up or wick away sweat. Saddle cloths and blankets made from natural fibres such as cotton, wool or felt are better than man-made fibres. Synthetic saddles should always have a saddle cloth underneath because they create a lot of heat. Western saddles must always be used in conjunction with a thick pad because the saddle itself does not have padded panels. A washable blanket is often used between the horse and the thick pad for ease of care.

Trying a saddle for fit

The issues involved with fitting are similar whatever the type of saddle. Poorly fitting saddles cause many problems with horses, including resistance in the horse when riding. Never compromise a horse's back by using a poorly fitting saddle. The saddle should first and foremost fit the horse because an uncomfortable or sore horse is potentially an unsafe horse. The next consideration is that the saddle should be comfortable for the rider, however, keep in mind that fitting the horse has the higher priority from a safety point of view.

Whenever possible get a professional to fit a saddle. A good saddlery shop should have a trained saddle fitter who will come out and fit a saddle to a horse, or sometimes it is possible to take the horse to the store. If buying a saddle from another source always try it on the horse first. This can be difficult when buying privately, but you can always ask if you may take a saddle to try on the proviso that you pay for it first and can take it back if it does not fit.

Remember that a horse changes shape as it loses or gains condition and as it develops muscle tone. For example a saddle that is fitted to a three-year-old horse will not necessarily fit that same horse after a year or two of work when it has developed muscle that will cause the back to widen. Another factor to keep in mind is that pannaled saddles, such as English or Stock saddles, need to be repacked every one to two years if used regularly.

To fit a saddle to a horse, stand the horse on a flat level surface. The saddle should initially be tried without a saddle cloth so that it can be checked thoroughly. If there is a need to keep the underneath of the saddle clean, place a thin towel between the saddle and the back – nothing thicker. The following checks should be carried out without a rider in place and then, as long as everything is correct, with a rider in place. A second person can help with checking the saddle with a rider in place by either sitting in the saddle or doing the checks from the ground while you are mounted.

- Before putting the saddle on the horse check that the gullet is no less than 8 cm wide down its length. Saddles with a narrow gullet put pressure on the spine and prevent the spine from lifting and bending when the horse is working.
- Check that the saddle clears the withers. If it doesn't, there is no point going any further
- The back of the saddle should not go further back than the horse's last rib. This can be felt with your fingers. A horse should never carry weight beyond this point because the back is not strong enough without the support of the ribs.
- The saddle tree (inside the saddle) should be the same width as the horse. To check this, run a hand between the saddle and the horse from the withers down the shoulder. Slightly wider is better than narrower.
- The panels should mirror the shape of the horse's back, putting even pressure along the muscles on either side of the spine. To check this, run a hand under the panels from front to back. If the panels are the wrong shape for the horse there will be more pressure in the middle, which will cause the saddle to rock, or less pressure in the middle, which will cause the saddle to bridge, putting too much pressure on the front and back of the saddle area.
- The horizontal line (of the seat) through the centre of the saddle should be parallel to the ground, not sloping up or down.
- The same checks should now be carried out with a rider mounted.

Now check that the saddle fits the individual rider. Deeper seats are tighter than flatter seats. Men are usually more comfortable in flatter saddles and women in deeper saddles. Western saddles come in different seat shapes. A pleasure saddle will generally place your legs more underneath you, and a cutting saddle will put your legs more forward. The best way to tell if the saddle fits and is comfortable is to ride in it. The saddle should not feel as if it tips you forwards or backwards. The horse should go happily in it without resistance and you should feel comfortable at all paces.

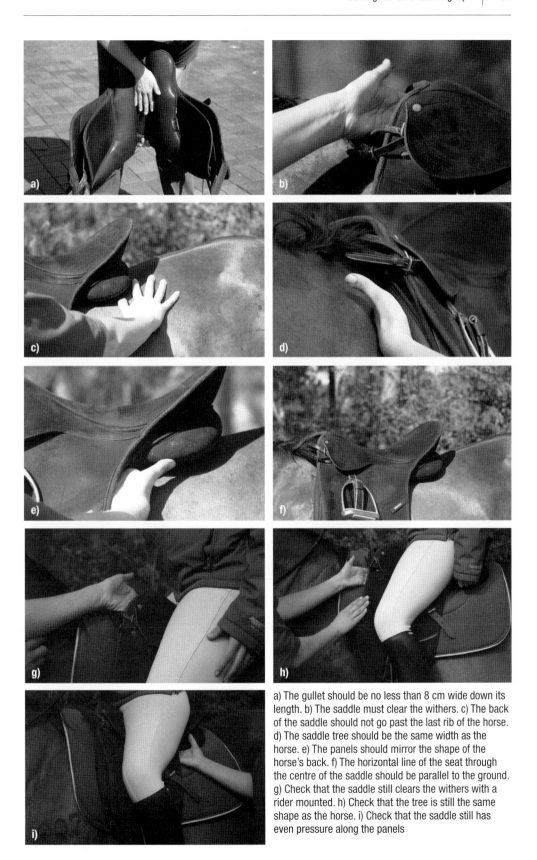

a) The gullet should be no less than 8 cm wide down its length. b) The saddle must clear the withers. c) The back of the saddle should not go past the last rib of the horse. d) The saddle tree should be the same width as the horse. e) The panels should mirror the shape of the horse's back. f) The horizontal line of the seat through the centre of the saddle should be parallel to the ground. g) Check that the saddle still clears the withers with a rider mounted. h) Check that the tree is still the same shape as the horse. i) Check that the saddle still has even pressure along the panels

Saddling for riding

Before putting on a saddle make sure the saddle area is clean. Even if the horse has been brushed always feel the area with the flat of a hand to check for any grit or sores (see p. 55).

Also check the saddle cloth or blanket before putting it on the horse. Feel it to make sure there are no protrusions or harmful objects such as sharp grass seeds, grit or insects.

A horse must be secured before saddling. It can be tied up safely or held by a handler. A horse should be trained to stand still for saddling irrespective of whether it is tied or held. To put a saddle on, follow these steps.

- Collect the saddle, checking that it is complete and that it is the correct saddle for that horse.
- Carry the saddle to the horse on one arm as shown, approaching the horse from the left.
- Stand on the horse's left (near) side and put the saddle cloth/blanket on the horse's back, then place the saddle on top of this. Aim to put them slightly too far forward then slide them back into place so that the hairs on the back are lying the right way. Make sure that the outside edge seems of the saddle cloth do not sit directly under the saddle as these will cause saddle sores.
- Cross to the other side of the horse and lower the girth, making sure that it is not twisted under the flap. Do not duck under the head and neck of a tethered horse (see p. 53).
- Girth up the saddle as soon as it is placed on the horse's back. Never leave a saddle on a horse without the girth being fastened as the horse may move suddenly and the saddle slip off. As well as damaging the saddle this can frighten the horse and cause it to panic.
- Cross back to the left side and fasten the girth. If the horse is to wear a martingale etc. that is attached via a loop, thread the girth through the loop. When a girth is first fastened it should be tight enough to prevent the saddle from slipping if the horse moves, but no tighter. Gradually tighten it as you prepare to ride. Tightening the girth completely straight away leads to a horse anticipating the tightness and either behaving defensively ('blowing out' or nipping/biting) or in some cases, panicking, which can result in it pulling back or showing other panic reactions.

Western saddles have two different fastening methods for the near side. One uses a buckle on the cinch, the second uses a cinch knot at the saddle. Western saddles with two cinches should always have the front one fastened first when saddling and the back one unfastened first when unsaddling, so that the saddle is never held on with only the slacker back strap. The back strap (rear cinch) holds the saddle in place when the rider is roping a cow: the cow is roped to the horn of the saddle and therefore the saddle can tip up at the back. A rear cinch should only be fastened snug around the barrel, never tight. The two cinches should be fastened together with a strap (under the belly), otherwise the rear cinch can slip back and cause the horse to buck.

Many people pull out the front legs one by one after they have fastened the girth. This was more important with string girths, to make sure no skin was trapped between the strings. It is less important or even unnecessary with smooth girths. If you choose to do

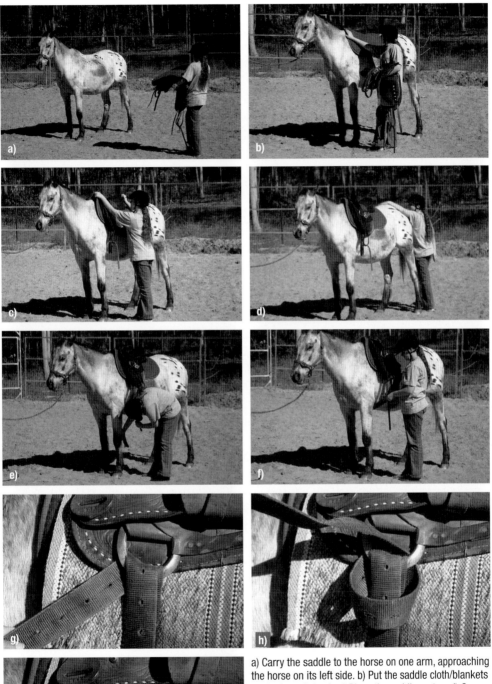

a) Carry the saddle to the horse on one arm, approaching the horse on its left side. b) Put the saddle cloth/blankets on the horse's back. c) Place the saddle on top. d) Cross to the other side of the horse and lower the girth. e–f) Cross back to the left side, reach under the horse and fasten the girth. Thread the girth through the martingale loop, if you're using one. g–i) How to tie a cinch knot on a Western saddle

a) Unfasten the girth. b) Lower it gently. c) Put your left arm up and under the channel of the saddle, pulling it over and off the horse. As it comes away from the horse catch the girth with your right hand and place it over the saddle

this, make sure you pull the leg gently forwards and lower it gently to the ground afterwards, otherwise it is possible to wrench the muscles in the girth area.

Removing a saddle

To remove a saddle, either tie the horse up safely or hold the horse. If the stirrups are not yet run up or crossed over the saddle, do this now. Unfasten any attachments (martingales etc.). If they are still attached and you attempt to pull the saddle off, the horse may panic when it feels the unfamiliar pull around the elbow. You will create the situation of having the saddle half on and half off without being able to pull the saddle clear of the panicking horse, which is a very dangerous position. Unfasten the girth and lower it gently. Put your left arm up and under the channel of the saddle, pulling the saddle over and off the horse. As it comes away from the horse catch the girth with your right hand and place it over the saddle. Then place the saddle in a safe area before seeing to the horse.

Problems with saddling

A common saddling problem is horses that bite when saddled. If this occurs you must take measures so that the horse can't bite anyone but also so that the horse learns a new behaviour. This can be a very difficult habit to break and it may not be possible to eradicate it, especially if the horse will be handled and saddled by different people in the future because consistent techniques are necessary if the horse is not to revert to its old ways.

As with all problem behaviours, make sure that the horse has no reason to misbehave. See if there is a cause that should be removed. Many horses exhibit this behaviour because they have been girthed up too roughly or quickly in the past. Removing the cause will not always remove the behaviour once it has become entrenched, but this is an important first step.

Tie the horse up short enough so that it cannot reach you with its teeth. Saddle the horse carefully. An elasticised girth is especially useful with this type of problem. Take even longer than normal to tighten the girth, ease it up until it is comfortably snug, then go up hole by hole until it is tight enough to mount. This process can take ten minutes or so. You can do other things during this time, such as making preparations to ride.

Wear and tear gear checks

Gear should be thoroughly checked for wear and tear periodically and quickly checked before each use. Cleaning is a good time to thoroughly check gear for wear and tear.

A thorough safety check involves the following.

- Rivets – check for loose or broken rivets on the saddle.
- Billets – check for damaged or loose billet hooks (hooks on an English bridle that secure the bit to the cheekpieces and the reins).
- Buckles – check all the buckles for rusty or bent buckle tongues. Check the leather that joins a buckle for cracking and wear.
- Strapping – check all strapping for cracking, tearing or stretching. In particular, check the areas where metal touches leather constantly in the same place (such as where the rein attaches to the bit, the stirrup leather joins the stirrup and the offside cinch knot on a western saddle) and for torn buckle holes that are tearing into each other (in particular the girth straps and stirrup leathers).
- Metal – check all metal parts (such as bits) for sharp edges. Check stirrup irons for rusting, cracking or bending. Check that the metal of the stirrup bars is not loose, corroded or cracked.
- Elastic – check any elastic parts for perishing (such as elastic on a girth).
- Webbing and rope – check for any frayed or rotten webbing/rope.
- Stitching – check for worn stitching, in particular the stirrup

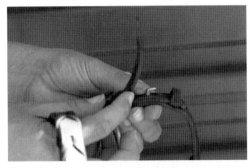
Check for damaged or loose billet hooks

Check for worn stitching by holding the two ends of the stirrup leathers and trying to pull them apart

Check for breaks in a tree

leathers, by holding the two ends and trying to pull them apart. Check the area where the girth straps join the saddle if they are stitched on as these can fray or become unstitched.

- Saddle – check for breaks or twists in the tree and for sharp edges or lumps in the panels. It is difficult for inexperienced people to check a tree so they should take the saddle to a professional saddler occasionally to get it checked out.

In addition to checking saddles and bridles, check all other gear that is used on the horse. If the gear is for driving, check the harness thoroughly. Broken driving harness is a common source of driving accidents.

Any items of gear that are found to be unsafe should be replaced or properly mended before being used again.

Maintenance of gear

All gear should be maintained on a regular basis; this reduces the chance of safety issues such as breakages and extends the life of gear. Maintenance is also an opportunity to give gear the thorough safety check described above. Good storage of gear is equally important.

Leather gear should be cleaned regularly. This involves taking it apart and washing the leather with warm (not hot) soapy water, making sure you remove the grease lumps that tend to form wherever leather comes into direct contact with the horse. An old toothbrush can assist with this! The leather should be dried with a cloth before you apply leather conditioner. Never dry leather with direct heat (such as in front of a heater) as this will make it dry out too quickly and crack. Conditioning the leather can be done with leather oil or leather conditioner. Too much oil can make leather too soft and stretchy.

Metal parts such as stirrup irons and bits should be washed with warm soapy water and dried with a cloth. It is not necessary to use metal polish on good-quality metal such as stainless steel. Never apply polish to the mouth parts of a bit. Bits and stirrups can be

Saddles stored properly

AHSE

put in a dishwasher occasionally for a good clean. Synthetic/plastic gear should be washed in warm soapy water from time to time.

Store gear in a clean, dry, well-ventilated area that is free from rodents. Remember that mice and rats like the taste of leather, so either have mouseproof storage or a pest eradication scheme! Leather is particularly prone to mould if it is stored in a damp humid area. All gear should be hung when not in use (not stored on the floor). Commercially produced gear hangers can be used; home-made ones work just as well.

7

Safe riding

Approximately 80% of injuries involving horses occur when riding them, the rest when handling horses and when near them on the ground. Some injuries occur when a rider is on the horse but the majority occur when the rider falls off. They range from minor to very serious, including serious neck and back injuries and even death from head injuries. Injuries are usually caused by the horse's legs and by contact with the ground or another object. It is quite possible to fall from a horse that is only walking and still sustain a serious injury. It all depends on the angle at which you hit the ground or other obstacles etc. Most accidents can be prevented and forewarned is forearmed.

This chapter does not aim to teach people how to ride or instructors how to teach. There are many excellent books on the subject, too numerous to list. There are many styles of riding with many variations within and between. All follow the basic pattern of balance and there are many more similarities than differences even though it may not seem that way to the casual observer.

Traditionally, certain cultures have evolved the different styles of riding, however whatever the chosen style the safety rules outlined throughout this book are equally important and these rules can and should be applied to any discipline.

Mounting

Mounting should take place in an area that does not have a roof or anything else overhead unless very high, such as indoor arena. The surface should be even and not slippery and the horse should not be positioned near any dangerous obstacles that could cause an injury to the rider or the horse if the rider or horse were to fall on them.

Safety checks before mounting

The following safety checks should be used to check gear before anyone mounts a horse. This gear-checking procedure should be used to check the gear of a horse that you are about to ride (either your own or someone else's) or used by instructors, parents or anyone that is responsible for other people who are about to ride (for the purpose of this section, the term 'assistant' describes anyone in charge of or helping someone to mount). If any item of gear is found to either not fit properly or to be unsafe in some other way it should be fixed before proceeding.

- The rider's gear should be checked, i.e. whether the helmet meets current standards, whether it is fastened securely, whether it fits, whether the rider is wearing the correct footwear. It is easy to forget, for example, to change footwear from workboots to riding boots before mounting (see p. 21).
- The horse's gear should be checked irrespective of who tacked the horse up. For example, check that all the straps are done up.
- If the gear does not belong to you and is therefore not subject to your own frequent wear and tear checks, quickly and thoroughly run through the procedure outlined earlier (p. 95). If you are not happy with the condition of the gear then do not mount or allow the rider to mount.
- Check that the bridle fits the horse, i.e. that the bit is in the correct place (not too high or too low), whether the noseband is fastened correctly, whether the browband is sitting in the correct place, i.e. not pulling the headpiece onto the ears (see p. 76).
- Check that the saddle has been put on properly and that the saddle cloth is in the correct position (see p. 84).
- Check the girth just before mounting. Check it again immediately after mounting.
- Before mounting, the stirrups should be let down. Check that the stirrup is the correct size for the boot (see p. 85). You can check this before the rider mounts by the rider putting their foot in the stirrup as if to mount. Alternatively, the stirrup can be removed from the saddle so that it can be tried on the foot before the rider mounts.
- Get a rough estimate of the correct length for the stirrup (leathers or fenders) by putting the knuckles of your hand against the stirrup bar and the bottom of the stirrup into your armpit. If it reaches the armpit the stirrups are roughly correct. They can be adjusted again when mounted.

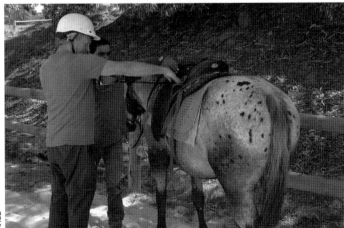

AHSE

Put the knuckles of your hand up against the stirrup bar and the bottom of the stirrup into your armpit. If the stirrup reaches your armpit the length is roughly correct

Common things that go wrong when mounting are described below.

- The girth is not tight enough and the saddle slips around. This can frighten the horse and is very dangerous as one of the stirrups will be hanging underneath the belly of the horse and the person mounting may fall under the horse. There are times when it is impossible for the girth to be tight enough for mounting from the ground, for example if the rider is heavy and/or awkward, if the horse is very tall and the rider small or if a very fat pony is completely round in shape. A mounting block should always be used in these cases.
- The rider digs their toe into the horse when mounting, causing the horse to jump forwards. In this case the rider needs to improve their technique so that this does not happen. If it happens a few times the horse will start to anticipate and will be unwilling to stand still. Again, using a mounting block helps to avoid this problem as the rider will find mounting easier.
- The rider lands heavily, which causes the horse to jump forwards or even to buck. Riders should lower themselves gently into the saddle rather than let their entire weight flop down. The rider must practise their technique until this no longer happens. Building up leg muscles helps to control this problem. If a horse bucks or dips when mounted without an obvious reason, check the back and saddle fit.
- The horse walks forward or around the rider as they are mounting. The horse should be trained to stand still before, during and after mounting (see p. 125). If you are assisting someone else to mount you must **not** hold the horse too tightly as a horse often lifts its head slightly to counterbalance the weight of the rider mounting. If it is held too tight it can rear or pull back. For the same reason a rider should **never** mount a horse that is tied up, because the horse can pull back.

A rider can mount from the ground (alone or assisted), receive a leg up from an assistant or mount from a mounting block (alone or assisted). **Inexperienced riders should not be expected to mount on their own**. An assistant will be required to help until they have practised and improved their technique.

Mounting from the ground is fine for agile and reasonably lightweight people if the horse is not too big. Receiving a leg up is fine if there is someone willing and able to do it and the rider is reasonably able-bodied. Mounting from a mounting block is usually the preferred and safest option for the sake of the rider, horse and saddle. If a rider is not able-bodied and would have to pull on the saddle to mount, they should definitely use a mounting block as otherwise the horse may be pulled over or the saddle may be damaged. Different mounting situations are described below.

Mounting from the ground without assistance

- Take the reins in your left hand with a light contact on each rein. If one rein is shorter than the other the horse will swing away from or towards you if it moves.
- Stand with your left shoulder parallel to the horse's left shoulder, which means that you are facing towards the tail.
- Take hold of the stirrup leather or fender and turn the inside (the side that hangs next to the saddle) to the back. Put your left foot in the stirrup pointing down then

reach up and over the saddle with your right hand to grasp the pommel or the skirt on the far side of the saddle.

- Hop two or three times on your right foot and spring up, turning as you go so that you are now facing the horse. You must be careful not to poke the horse's ribs with your foot while mounting.
- Don't throw your right leg over the horse until your body is straight up, otherwise you might fall backwards.
- Slow your motion so that you lower yourself gently into the saddle rather than land heavily.
- Once in the saddle, put your right foot into the right stirrup iron. You may need to lean down slightly to get hold of the stirrup leather/fender and turn the inside to the front (the other one will now face the front after you turned when mounting). This is the correct and most comfortable way for the leathers/fenders to hang.
- You may need to briefly shift a little extra weight into the right stirrup to straighten the saddle after mounting.

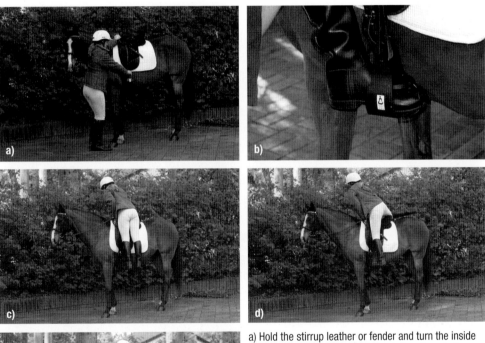

a) Hold the stirrup leather or fender and turn the inside to the back. b) Put your left foot pointing down in the stirrup. c) Spring up, turning as you go so that you face the horse. d) Don't throw your right leg over the horse until your body is straight up. e) Slow your motion so that you lower yourself gently into the saddle

Stirrup leather facing the correct way.

Fender facing the correct way

Altering stirrup length when mounted

For general riding the stirrups should be level with your ankle when your foot is out of the stirrup. An unassisted rider can alter the stirrups on an English saddle by keeping both feet in the stirrups, putting the reins into one hand and using the other hand to alter the stirrup while still looking forwards. Western or Australian stock saddles usually require an assistant on the ground to alter the stirrups, or you should dismount, alter them, then remount.

A rider altering the stirrups while mounted

Tightening the girth when mounted

The girth now needs to be checked and if necessary tightened because your weight will have caused the saddle to sit closer to the back and the girth may have become loose. If you have an assistant that person can check and alter the girth if necessary.

A mounted rider can tighten the girth of an English or Australian stock saddle. Your feet should remain in the stirrups when tightening the girth. If the saddle has long girth straps (and therefore a short girth) hold the reins in one hand, lean forward and down the side of the horse, put your other hand around the girth strap, pull up and put the tongue of the buckle into the hole with your index finger. If the saddle has short girth straps put your leg forwards, again keeping both feet in the stirrups, hold the reins in the opposite hand, lift the saddle flap and hold it up with the thumb of the hand that is holding the reins. Your other hand can now be used to tighten the girth.

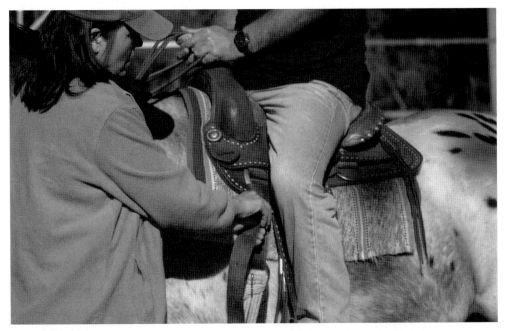

An assistant tightening the girth

A rider tightening a short girth (long girth straps) while mounted

An assistant can help the rider by adjusting the stirrups to the right length and can check and tighten the girth if necessary

The girth of a western saddle should not be tightened by an unassisted rider while mounted. The rider will need to do the girth up a bit tighter before mounting or dismount to do it.

After about ten minutes of riding either the rider (if capable) or an assistant should check the girth again.

Assisting to mount from the ground

When you are assisting a rider to mount from the ground, you can help by holding the horse if necessary (in addition to the rider holding the reins as they mount). You can do this either by holding the halter (if it has been left under the bridle) or the cheekpiece of the bridle. Do not hold the horse tightly. You can then help the rider to alter and to place their feet in the stirrups and can check and tighten the girth if necessary.

Assisting to mount with a leg up

A rider can be legged up on to a horse by another person. Traditionally this is done by holding the bent leg of the rider around the calf and lifting as they jump up. However, this can cause back injuries to the assistant, especially if they assist frequently.

An alternative technique, which can be used if you are an assistant, is as follows.

- The rider holds the reins.
- The stirrup leather is lengthened if the rider is having difficulty reaching the stirrup.
- The rider stands facing towards the horse, with the assistant by the horse's left shoulder.
- The rider places their left foot in the stirrup and you (keeping your back straight and bending your knees) support the heel of their left foot.
- The rider springs up with you helping the foot to stay in the stirrup and (keeping your back straight and your knees slightly bent) lifting the rider as they spring, if necessary. This gives the foot a stable platform and keeps the rider's left toe away from the horse's side.
- Once mounted the rider takes a rein in each hand so that they have control of the horse.
- You can then help the rider by adjusting the stirrups to the right length and check and tighten the girth if necessary.

Put your hands under the rider's heel.

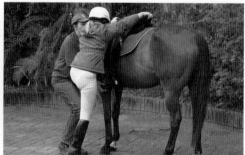

The rider springs up while you lift with your knees bent and your back straight.

The rider throws their leg over and lowers themselves gently into the saddle

This technique is especially useful for a small person with a large horse or when you are helping beginners to mount the first few times.

When assisting a rider it is appropriate to touch a rider only below the elbow and below the knee. Emergencies require individual judgement.

Mounting with a mounting block

A mounting block should be solid and safe. Old milk crates are commonly used as mounting blocks and are dangerous. They are usually made from plastic that is not UV stabilised and they tend to shatter or tip over without warning. A good mounting block should have a wide base, preferable wider than the top, for stability.

Where riders who are not able-bodied have to be mounted on a regular basis (such as in a commercial riding centre or riding for the disabled (RDA)) a permanent block or even ramp may be necessary.

A purpose-made mounting block

A rider mounting from a block

A purpose-built ramp used for riding for the disabled

The technique for mounting with a mounting block is the same as when you are mounting from the ground. You still hold the reins and put your left foot in the stirrup to mount. The only difference is that the mounting block means you start the procedure from a higher level and therefore don't have to jump or pull your weight up. Assistants are still necessary if the rider is a beginner or is not able-bodied.

Dismounting

When inexperienced riders are dismounting an assistant should stand by the head of the horse to assist if necessary. The assistant should check that the right foot comes out of the stirrup before the rider begins to lift the leg to dismount.

There are several things that can go wrong while you are dismounting.

- The horse walks off as you are dismounting. This can cause you to lose balance and fall. A horse should be trained to stand still for mounting and dismounting (see p. 125).
- You can get hooked up (by your clothing) on the front of the saddle. This mainly applies to western saddles and sometimes to Australian stock saddles. Always tuck loose clothing in.
- It is possible for your right leg to get caught on the back of the saddle if the leg does not clear the back of the saddle sufficiently. This is common if you are wearing long riding boots or aren't supple enough to swing your leg clear of the saddle. Practise your technique until it improves. For example, if your leg is not clearing the back of the saddle you are probably not leaning forwards enough or pushing up with your hands enough.

- The right stirrup can catch on the foot as you swing your leg back instead of coming off the foot. This is particularly a problem with lighter weight stirrups (it is not usually a problem with heavier western stirrups). This leaves you in a very dangerous position. Always make sure the stirrup has come off your right foot before swinging your leg back.
- The horse can throw its head back as you lean forward, hitting you in the face. Some horses have a habit of doing this and should be retrained by an experienced horse person to lower the head as the rider dismounts. You should always put your head to the side of the horse rather than directly forward when dismounting.

The correct way to dismount from an English saddle is to put both reins into your left hand and at the same time taking both feet out of the stirrups. Lean forward from the

a) Dismount from an English saddle by putting both reins into your left hand while taking both feet out of the stirrups. b) Lean forward from the waist, putting your left hand on the pommel and your right hand just above the stirrup bar (on the skirt). Swing both legs back, swinging your right leg clear of the back of the saddle, and at the same time move your upper body to the right across the saddle. c) Bring both legs together and slide down to the ground on the left side of the horse landing with your knees slightly bent

waist, putting your left hand on the pommel and your right hand just above the stirrup bar (on the right hand skirt) and simultaneously swing both legs back, swinging your right leg clear of the back (cantle) of the saddle. At the same time your upper body moves to the right across the saddle. You then bring both legs together and slide down to the ground on the left side of the horse landing with your knees slightly bent.

Beginners should dismount from western saddles and Australian stock saddles in the same way, putting their right hand on the horn of the western saddle or the raised areas at the front of a stock saddle. They can't lean as far forward, but they can use their right hand to push up with. The right hand helps to prevent their clothing from getting caught on any projections and there are two contact points for safety (left hand with short reins and right hand for stability).

Beginners should always dismount as described above and commercial riding centres (who deal mainly with beginners) should make sure that clients dismount safely.

Experienced western riders and Australian stock riders customarily dismount by keeping their left foot in the stirrup as they swing their right leg over the back of the saddle and step down. One argument for this is that it reduces the chance of clothing getting caught on any projections. The problem is that if the horse moves or if the right foot slips when the rider steps down they can lose balance and even be dragged if the left foot does not come out of the stirrup. Another way of dismounting these types of saddle is to take the right foot out of the stirrup and swing the right leg over the back of the saddle, bringing it to meet the left foot. The rider then quits the left stirrup and slides down to the ground.

Dismount from an Australian stock or western saddle the same way you dismount from an English saddle

You should *never* throw your leg forward over the horse's neck to dismount. This is a very unsafe practice for several reasons.

- You don't have control of the horse as your leg passes over your hands.
- You may fall backwards over the other side of the horse if it moves suddenly.
- You can accidentally kick the horse if it moves its head suddenly, which it may do when it sees the your leg cross over its head.

Once you have dismounted you should run up the stirrups or cross them over the saddle (if using a western or Australian stock saddle). Stirrups should always be put up out of the way when not in use. Hanging stirrups can catch on projections which can frighten the horse and damage the saddle and a horse may get a back foot caught in one if it kicks forwards at a fly.

The girth should be loosened a couple of holes if the horse is not going to be remounted immediately. Hold the horse with your right hand holding both reins behind the bit and your left hand holding the remainder of the reins across the palm of your left hand.

Loosen the girth a couple of holes if the horse is not going to be remounted immediately. Your right hand holds both reins behind the bit and your left hand holds the remainder of the reins across the palm

Riding

As stated earlier, there are many variations of riding ranging from western to dressage, from polo to showjumping and from trail riding to show riding. These are just some of the numerous things that can be done with a horse. The safety issues, however, remain the same. When riding, you should always follow these precautions.

- Always look ahead and plan ahead when riding and use your peripheral vision to see potential problems before they occur.
- Never ride a horse that you can't control on the ground. If a horse is unruly on the ground it is demonstrating a lack of training that must be addressed. An experienced rider may be able to control the horse when mounted but they would also be able to control it on the ground. It is a myth that a rider has more control mounted – this is only true if you are a very experienced rider.
- Never ride a horse on which you feel out of control. If you get a gut feeling that you are in a dangerous situation you should dismount.
- Don't trot before you can walk, canter before you can trot (including rising and sitting without losing balance, and controlling the horse on the straight and around turns), or jump before you can canter. You should have a secure seat at each pace before proceeding to the next one.
- Only ride in an enclosed area such as a round yard or an arena until you can control the movements of the horse. At the absolute minimum you must be able to make the horse stop, move on and turn.
- Never chew gum, smoke, eat or drink while riding. If you need to drink water you should dismount then drink.
- Never ride two up on a horse (apart from when taking part in organised 'vaulting' where the horse is trained and conditioned for such work and the situation is controlled). Riding two up is bad for the horse's back and is unsafe for many reasons. The second rider is sitting on a sensitive area that can cause some horses to buck, their legs make contact with the horse much further back than a single rider, if they unbalance they tend to pull the other rider with them and they have no control of the horse.
- Never hold someone in front (such as a small child) as this is a very dangerous practice. A horse cannot be controlled properly in this position. Putting a child up behind the saddle is also dangerous for the same reasons outlined above.
- Never allow a horse to graze when mounted. The horse can get a leg over a rein, step on a rein and pull back or, if the horse lifts its head suddenly and moves forward quickly, it can go very fast before you have time to gather the reins.
- Never ride in an area where there are loose horses, such as a paddock or an arena. This is a common cause of broken legs – you can be kicked by a loose horse that was actually kicking at your horse.
- Avoid riding in an area that opens onto a busy road with no barrier.
- Never ride near steel posts (star pickets) as they can cause serious injury if you fall on them.
- Don't ride on footpaths unless it is legal to do so. Stick to designated paths in wooded and bush areas and respect private property.

An in-control rider

- Always insist that the horse walks quietly for the last kilometre or so home, irrespective of the speed at which you have been riding. This also cools the horse down gradually. Likewise, don't canter or gallop up every hill or set off into a canter every time the horse steps onto a grassed (rather than harder) surface, otherwise the horse will begin to anticipate that hills and grassed surfaces mean go fast.
- Jump only when other people or riders are present.

Mobile phones

Only carry a mobile phone if you are going out for a ride. The horse should be habituated to ringing and beeping phone noises before you ride with it. Alternatively, switch the phone to vibrate only or switch it off altogether. It is possible to buy a pouch to fasten the phone to the outside of your upper arm, which is a relatively safe place to carry it. There is no point in fastening the phone to the saddle because if you fall off the horse may run away and leave you horseless and phoneless! You should dismount to answer any calls.

Riding on the roads

Riding on the roads is an increasingly necessary evil as safe areas to ride become fewer and farther between.

Common situations that occur when riding on the roads include those listed here.

- Vehicles can pass too fast and/or too close. Most drivers are not aware that horses can and do shy.
- Rubbish blowing around can cause the horse to shy into traffic.

- Loose dogs may run out of open property gates to bark/chase/nip or bite.
- Scary noises or sights on the roadside or a property can cause a horse to shy into traffic. This can include such everyday occurrences as children playing, people mowing lawns etc.
- Car drivers or passengers may deliberately try to upset the horse as they go past.

If you ride on the roads you must adhere to the road rules, in particular the ones that apply to animals on the road. Therefore you should be familiar with them before venturing out. Check with your transport department. In addition to laws about riding on the roads there are certain precautions you should take.

- Learn and use hand signals, but be aware that many drivers may not understand them. Drivers are given little if any tuition about what to do when they meet animals on the road. Keep this in mind and ride defensively.
- Children should not be allowed to ride on roads unaccompanied and any child should be trained in using the roads before being allowed to ride on them. A child is in more danger on the roads than an adult because they have less experience and knowledge of roads and traffic in general.
- You should only ride on the roads with a horse that is well-trained, that you can control and that is safe in traffic. Horses that shy frequently should not be ridden on the roads. A well-trained horse will stop, go and move over from your leg when asked. All these skills are imperative when riding on the roads.
- Before venturing onto the roads habituate the horse (see habituation p. 128) to as many scary objects and situations as possible. Try to set up a situation whereby the horse experiences traffic without being on a road. For example, taking a horse to a show (even if not competing) can habituate the horse to many scary situations including traffic.
- If you are an inexperienced rider you should get a more experienced person to accompany you when you first ride on the roads. The experienced horse person should ride a quiet well-trained horse that will give you and your horse confidence. If you're inexperienced, don't take an inexperienced horse out onto the roads for the first time. This class of horse should be introduced to the roads by an experienced rider.
- When riding with another person riding abreast causes drivers to slow down and pass wide. Ride no more than 1.5 m apart. Have the least experienced horse on the inside (kerbside). Only ride abreast if the horses are accustomed to one another. Keep their heads level to reduce the risk of kicking.
- Take care when riding over bridges. Ride on the road, not the footpath unless signed otherwise and be aware that the rails are usually designed for pedestrians not horses – a horse can easily fall over the side if it shies into the rails. If you're not confident about your ability to control a particular horse on a bridge you should dismount and lead it on the footpath.
- Avoid riding on high-speed roads, especially where the verge is narrow.
- Beware when riding on road markings such as paint marks as they are slippery, especially for a shod horse.

- Always make yourself as visible as possible whatever the time of day and avoid riding after dark. If riding at night is unavoidable, use lights and as much fluorescent gear as possible such as reflective horse boots, a reflective vest and a reflective stripe on your helmet.
- Always be courteous to drivers who slow down and/or move over. Smile or nod your thanks and then hopefully they will do it again the next time that they pass a horse.

The average driver does not understand that a horse is likely to shy at a plastic bag and move in front of a vehicle. They do not know how unpredictable a horse can be. Consider starting a campaign to educate drivers in your area. You might be able to do it through a state or national horse organisation or you may have to get together with like-minded people. Local radio may be willing to do public service announcements about horses on the road. Flyers that educate drivers can be placed in places such as service stations where they are likely to be picked up.

Riders on the roads must be very careful

Recommended further reading
Vicroads has a website with a page on horses and traffic. Go to www.vicroads.vic.gov.au then road safety then horses and traffic.

Riding alone
Going out for a ride alone with just your horse for company can be wonderful but you must be in control, which takes a certain level of skill. If you are an inexperienced rider it is not safe to ride out alone because if something goes wrong there is no one else to help you, and horses tend to be more anxious when on their own. In fact horses vary in how well they behave when alone. An experienced rider can often overcome this anxiety by giving clear signals to the horse so that it feels secure, but inexperienced riders of course

find this difficult. Having someone else there means that the horses can give each other confidence and in case of an accident there is someone to assist.

When you are more experienced, if you do decide to ride out alone tell someone where you are going and what time you expect to be back. Taking a mobile phone will give you a better chance of getting help if you need it.

Riding in groups

When riding with a group of friends or as part of an organised group (club) ride, certain procedures must be used so that everyone is safe. This section assumes a group ride of one or two hours' duration in relatively populated areas. Riding in wilderness areas for a longer time is outside the scope of this book, although the same basic principles of preparation and awareness still apply.

Potential dangers involved with group rides are that horses can become excited and unruly, people and horses can get kicked and riders can get left behind. Horses that are part of a commercial establishment and that do trail riding/trekking get accustomed to being in the same group, are selected for their temperament and the ride is managed by guides. When a group of friends get together none of these factors are present therefore it is a potentially more dangerous situation.

When riding as part of a group spend a little time before the ride planning and discussing how to keep safe. Designate an experienced rider as the leader to ride at the front, and another experienced rider to ride at the back. The following procedures will help to keep the ride safe.

- Check the weather and the route and make sure the riders have food, water and appropriate clothing. Don't hesitate to cancel if a thunderstorm is possible or the weather is inappropriate.
- Make sure the horses are fit and well and used to the work that they will be doing. For example, taking an unfit horse or a young one on a long ride for its first outing could cause enormous difficulties.
- If possible, check the route beforehand to make sure it is safe and that the length of the ride will be appropriate for the riders and the horses.
- Take a first aid kit (someone must know how to use it) and a mobile phone. It is also a good idea to have a halter and leadrope for each horse (this can be worn under the bridle).
- Inform a responsible person of your route and what time you should return. Sometimes mobile phones are unreliable, and you need someone to check that you are back safely.
- Ensure that everyone has mounted safely and checked their girths before moving off. All riders should be ready to mount at the same time, not some still grooming as others are mounted, for example. Everyone should mount in a safe manner, not too close to the other horses.
- The lead rider should check that everyone is ready before moving off and changing pace. No rider should pass the leader and riders should ask permission before passing another horse and rider. Pass at a steady pace. Never overtake each other when cantering as this can incite the horses to race each other.

- Wait for any stragglers or wait until all the horses in the group have drunk their fill at a water point before moving on. When crossing water be patient and wait for those that are having trouble. If anyone needs to dismount or is having any difficulty the whole ride should halt and wait.
- Keep to a steady speed in especially in open areas as horses can get out of control quicker in an open space than on a track.
- **Ride at the pace of the most inexperienced or nervous rider.** With a large group it is safer to keep to a walk because horses get excited when moving fast in a group. If you do go faster do so uphill as it is easier to keep control. Never canter downhill unless every group member is experienced and confident.
- In hilly country, wait until all the riders are at the bottom of the previous hill if you are going to trot or canter up the next hill.
- Be courteous to pedestrians and cyclists. Some people are very frightened of horses, especially a large group of them.
- The whole ride should wait while gates are shut, to ensure the person shutting the gate can control their horse and does not need to race to catch up.
- The whole ride should maintain safe distances. There should be one horse's distance between horses (each rider should be able to see the heels of the horse in front) and individual riders should not allow their horse to lag behind and cause large gaps.
- Only ride abreast in pairs if the horses are accustomed to one another. Ride with the horses' heads level to minimise kicking danger.
- The rider at the back is responsible for making sure that no one gets left behind.

If riding abreast, keep the heads of the horses level to reduce the chances of kicking.

- If a rider loses control and gets out in front of the ride do not chase after them. Follow at a steady pace. By chasing them the rest of the group may get out of control and the horse being 'chased' will only run faster when it is aware of a group of horses running after it. As far as the horse is concerned it is part of herd that is galloping away from something (see p. 8).
- If you suspect that your horse kicks, ride at the back. Simply putting a red ribbon on its tail is not good enough.
- If crossing a road, cross as one group. **Never** have some horses on one side of the road while some are on the other.
- Look after each other and the ride will be fun for everyone. Riding in a group is a social occasion therefore even though you may not be able to ride as fast as you like it is an opportunity to ride with like-minded people.

Riding in an arena

When you are riding in an arena with other people and everyone is working on their individual training, certain rules should be followed to avoid collisions and kicking. It is a good idea to post a list of rules on the entrance so that people are aware of them before entering the arena.

Some suggestions for rules are given here.

- Before entering the arena, check that the entrance is clear. The person wishing to enter should shout 'Door free?' and wait for a reply before entering.
- Riders must dismount to enter and leave.
- Gates/doors must be closed at all times.
- No clothing, drink bottles etc. may be placed on the ground or the fence.
- Riders should pass left shoulder to left shoulder. This avoids head-on collisions.
- Riders should maintain one horse's distance when riding in the same direction. To check this, a rider should be able to see the heels of the horse in front between their horse's ears.
- No horses to be tied or loose in the arena when people are riding or lunging.
- No lunging in the arena when people are riding.
- Jumps must be put away after use. In a jumping arena where the stands are to be left out, jump cups that are not holding poles must be put away safely.
- Riders must always check behind them before spinning or backing.
- Older and more experienced riders should look out for younger and less experienced riders.
- Riders travelling at a slower speed should travel on the inside track, allowing those that are travelling faster to overtake on the outside track.
- Riders must check verbally that it is okay to pass before overtaking.
- If a rider needs to dismount, they must ride to the centre of the arena and dismount there.
- If a rider knows that their horse is a kicker they should not ride in the arena with other people.
- When leaving the arena shout 'Door free?' and wait for confirmation before opening the gate/door.
- After riding, riders must pick up any manure.

Riding at a show or event

Taking a horse to a show or an event of some kind can either be lots of fun or a nightmare, depending on the horse's behaviour. A different environment is always exciting or frightening to a horse and when that is coupled with lots of horses and numerous other stimuli such as different types of animals, noisy machinery, lots of people etc. it is not surprising that many horses misbehave at a show. As with many situations, preparation is the key to a safe and enjoyable day. You must be able to control and therefore manoeuvre your horse forwards, backwards and sideways on the ground and in the saddle (if riding) (see p. 122).

If the horse is young and/or has never been to a show, plan to take it for the day without entering in any classes. For many horses, doing this once will let them relax enough for you to enter them at the next show. If not, keep taking the horse out and about until it is calmer. It is also a good idea to enter only led/in-hand classes at the first show. As well as relieving the pressure on the horse it also relieves the pressure on you, which is normally transferred to the horse.

There are potential hazards at shows and events.

- Other horses or people may get too close to the back of your horse. Keep a close eye and be firm about not allowing people to do this.
- Small children run around and there are families with prams/pushers and balloons etc. Families often attend events to watch but may not understand horse behaviour. They may let children get far too close without realising the dangers. Horses that have never seen a pram can be initially frightened of them.
- Loose animals – inevitably animals get loose at a show which can result in many other animals becoming excited. Loose animals can panic and charge through ring ropes and groups of people, creating a potential for injury to people and horses.
- Strange animals – there are classes for many different types of animals at agricultural shows. Many horses are frightened of donkeys, for example, if they have never seen one before so it is a good idea to your horse to donkeys before the day of the show. Even a horse-drawn vehicle can frighten a horse that has not seen one before.
- Larger shows can also have many other potentially frightening events such as people skydiving into the next ring, helicopter rides etc. These things are difficult to habituate a horse to before the day, however, the more a horse has seen and done before going to a show the easier it will accept new situations.

Falling off

If you ride frequently the chances are that you will fall off at some point. This does **not** mean that you have to fall a certain number of times to become a rider, which is a common myth. Aim to never fall off, but be prepared if it happens.

If you find yourself falling off try to tuck your head in and roll if possible. Even though it may be difficult, it is important to relax as you fall. The tenser you are the more likely you are to get injured. Always let go of the reins when you fall off. People who keep hold of the reins have been dragged and seriously injured when they would have been fine had they let go.

Learning to fall properly is a good idea. You can do this by attending martial arts or gymnastics classes (see p. 17).

You should only remount (or be allowed to remount) if you have no injuries and did not bang your head when landing. Even so, it is a good idea to wait a few minutes in case shock or after-effects appear. Even if you're not injured and did not bang your head, you should never be pressured into getting back on if you don't want to.

If you hit your head you must not remount. Any bang to the head should be monitored and checked by a professional if there are any symptoms of concussion (these can occur even days after the impact). It is possible to get a concussion injury when wearing a helmet – the helmet has reduced what may have been a fatal hit to the head. A helmet that has been knocked must be destroyed (see p. 19).

Any injuries can be **compounded** if you remount with an injury and fall off again. The level of damage from the first injury can be vastly increased by the second injury.

Improving skills

Any experienced horse person will tell you that learning to ride and improving other horsemanship skills is a never-ending process. Even Olympic riders have coaches. Anyone who says after just a few riding lessons that they can now ride, is demonstrating a common misconception that arises from lack of knowledge.

Lessons are not only for people who want to compete, but for people who want to ride safely and to keep their horse comfortable. A comfortable horse is a safer horse. A rider who is not balanced, who has to grip with the legs or hang on with the reins will make a horse uncomfortable. A responsible rider wants to ride to the best of their ability. One of the most important things you can acquire is a secure independent seat. This means that you can stay with the horse at walk, trot and canter, up and down hills without losing balance and without having to grip with your legs or hang on with the reins.

There are various ways to improve your skills and a combination usually works best. Books, videos or DVDs, attending private or group lessons at a riding school, having an

Having regular lessons is part of learning to ride and improving your riding skills

Riding out in the open over different terrain helps to develop a secure and independent seat

instructor teach you at home or attending clinics with your horse are some options. You can 'fence-sit' at a clinic and still learn a lot about handling and riding even without owning a horse.

Some universities and TAFEs offer courses in Horse Studies for people who are interested in making horses their career.

Learning to ride a horse and learning horsemanship skills is a long-term prospect; it takes time to be able to ride and handle horses well. Any experienced horse person will say that the more you learn, the more you realise there is to learn! Learning is fun at every stage, be patient and enjoy the journey.

Choosing a place to learn to ride

If looking at a riding centre or school with a view to yourself or your children having lessons, there are certain things that you should check out first. Go for a visit and watch a lesson.

- Is the site accredited and are the instructors qualified (see p. 164)?
- Is the place tidy and neat? This is irrespective of how fancy the facilities are.
- Do they have a list of rules posted (about behaviour around horses, safety issues etc)?
- Is there a securely fenced arena?
- Is the tack in good repair?
- Are the horses in good condition?
- Are the instructors dressed appropriately?
- Does the place insist on helmets and does the instructor wear one when riding?
- Does the centre require riders' medical and contact details?

- Do the instructors emphasise safety?
- How many people are there in a group lesson? There should be a maximum of eight per instructor.
- Do they do private lessons for beginners?
- Do the instructors have first aid training?
- Is there communications and first aid equipment nearby for emergencies?
- Do they have public liability insurance from a recognised insurer?

Talk to clients to see if they are happy with the service.

Common riding problems

When horses do not behave the way we want it is too easy to blame the horse and even assume that it is misbehaving on purpose. This is a common misconception about horses even among experienced horse people. Misbehaving deliberately requires reasoning, which horses cannot do. Horses learn that one type of behaviour elicits a certain response and repeat that behaviour if it profits them.

Common riding problems with horses involve behaviours such as baulking when asked to go somewhere, bucking or rearing. Problems are generally due to one or a combination of the following reasons.

- Lack of correct training.
- The horse is uncomfortable or in pain.
- The horse has too much or not enough energy.
- The horse does not understand what it is being asked to do.
- The horse is physically incapable of doing what is asked.
- The horse has learned to activate the flight response in certain situations.
- The rider is not experienced enough to handle/ride the horse.

Therefore, the beginning of any solution is to address these issues and make the necessary adjustments. Often simply implementing correct training and horse management is enough to address mild problems, however, some entrenched and dangerous problems need the skills of an experienced trainer if they are not to get even worse. Problems need to be sorted out as soon as possible before they become entrenched behaviour. Never be afraid to admit defeat and engage a professional if you do not have the experience to deal with problem behaviours. If the horse is sent to a professional trainer you will also need to learn how to prevent the problem from recurring, so make sure the trainer shows you the proper technique before you get the horse back. In fact, a trainer that is prepared to work with you and the horse may be more useful than one that will only work on the horse.

The first thing to do when looking for the root of a problem is to run through a checklist of possible causes.

- Has the horse been trained to do the task that you are asking it (see p. 122)? Has the horse learned this behaviour as a result of poor handling/training in the past?
- Check that the gear fits the horse properly. If gear is too tight and pinching or rubbing the horse, sooner or later the horse will object. For example, poorly fitting saddles are a common cause of problem behaviour (see p. 75).

- Check that the horse is healthy and sound. The horse should be in good condition (not too fat or too thin). It should be regularly wormed, have its teeth checked etc. Any signs of sickness should be investigated and a vet called in if necessary (see p. 45).
- Is the horse overly confined with not enough exercise? A horse that spends too long standing still will be keen to move and may play up simply through having too much energy (see below).
- Are the expectations realistic for this particular horse? The horse may be physically incapable of doing as asked. For example, it may be the wrong body shape (wrong conformation) or too young or too old (see p. 124).
- Is the rider 'overmounted'? The horse may simply be too much for the rider's present skill level. How good are their techniques and could they be improved? Everybody has room for improvement and should attend lessons or clinics frequently or from time to time to improve their skills (see p. 118).

Exercise levels

Horses need regular exercise, especially if they are confined to stables or yards. Horses that are confined need more exercise than those that are not because they are forced to stand still for long periods of time. Horses that are confined and under-exercised can understandably become difficult to handle and ride when they do get out. Horses are designed to move around a large area and forage for a large part of each day. Unless a horse can be turned out into a large area for several hours a day it should be exercised every day. Preferably, a confined horse should receive both turn-out time and structured exercise, such as ridden or groundwork.

Even horses that live in paddocks full-time should be handled or ridden on a regular basis. This does not mean every day, but the horse should be handled or ridden regularly. This maintains the status quo between a handler/rider and a horse.

Horses vary in how often they have to be ridden in order to maintain good behaviour. For example, some horses can be ridden infrequently with no discernible difference in their behaviour each time they are ridden; other horses become 'twitchy' if they are not ridden very frequently. A twitchy horse is unsuitable for an inexperienced rider and can be a problem and nuisance even for an experienced rider if it needs a retraining session every time it is brought back into work.

If a horse can only be ridden infrequently (for many people this is the only option) you should consider doing some groundwork exercises before mounting to check that the horse still yields to pressure and get it to focus on you (see p. 125). Groundwork exercises are invaluable for this and have other advantages such as warming up and suppling the horse. If you are not comfortable and confident about handling a horse on the ground you should not mount it. Find an instructor/trainer or attend a clinic that teaches you how to move a horse around on the ground (see p. 118).

8

Training for safe horses

This chapter looks at training a horse to be safer both on the ground and under saddle. There are a multitude of different training methods, however, the most effective are actually very similar. They may use different terminology but they tend to use pressure (the rider applies a leg aid) and the removal of pressure (the rider removes the leg pressure when the horse responds correctly). It is the removal of pressure that trains the horse.

Whenever a person interacts with a horse they are shaping its behaviour (training) in either a positive or a negative way. A horse only learns bad behaviours when a handler or a rider allows it to.

There are certain minimum handling requirements that all domestic horses should meet so that they can be handled safely in various situations and by people other than the owner. There may be situations such as a horse getting out onto the road, or an impending fire meaning that a horse has to be evacuated quickly. In both these situations a horse may, out of necessity, have to be handled by someone other than the owner. All domestic horses (including young horses) should be able to be caught and led, tied up, handled and loaded onto a horse transporter without fuss by any reasonably competent horse person.

It is essential for a horse to be well-trained. A quiet horse and a trained horse are not the necessarily the same thing. A quiet horse may be quiet either because it is well-trained or it may simply be unresponsive to the sorts of things that upset the average horse. This kind of horse will still become dangerous when pushed out of its comfort zone. A trained horse is quiet because it has been trained to be quiet. Its training will usually override its natural responses in a highly stimulating situation because a well-trained horse can be kept under control much more easily than an untrained horse.

Horses tend to get labelled – common examples are mean, dirty, lazy and stubborn or, at the other end of the scale, sweet, kind or willing. In fact the horse is none of these things; it is a horse and should not be given human emotions. Understanding the brain of the horse and accepting that it is different from our own is not in any way demeaning to the horse; in fact treating a horse like a human is demeaning as horses are horses not people! Humans tend to do this with all animals and even with inanimate objects (such as their car).

Being anthropomorphic leads to putting inappropriate expectations on a horse (see p. 9). This can lead to punishment if a trainer believes that the horse 'knows' when it has done wrong. A horse that misbehaves does so because of incorrect or incomplete training, not because it decides to be bad. This is very difficult for people to accept because it goes against what we are usually told about horses from when we first learn to ride. Recognising that a horse is a horse means that it can be trained or retrained without hampering it with labels that cloud our image of that particular horse.

Consistency is extremely important but is a quality that often has to be taught to people. Animals cannot be expected to understand a person's different moods from day to day. Inconsistency leads to confusion in a horse, and later maybe to dangerous behaviour.

Punishment is **not** an effective training tool yet it is often used by people who do not fully understand training methodology. Good training is where the horse is set up to do the right thing and rewarded for it, rather than punishing the horse for doing the wrong thing and not offering an alternative. Punishment may tell a horse what not to do but it does not tell the horse what to do.

All horses require training to some extent throughout life. You can buy a well-trained horse but if your cues are inconsistent the horse will eventually learn different and usually less desirable behaviours. As training is an ongoing process, if you take on the responsibility of horse ownership you must also take on the responsibility of trainer.

Horse/human relationships

People talk about 'bonding' with a horse, which implies that horses bond with humans and vice versa. It is more accurate to say that horses, through the process of good training, become calm, quiet and well-behaved which leads to them becoming relaxed around humans and giving the correct response to cues in most (if not all) situations. Training a horse to such a level requires good horsemanship. Good horsemanship can be learned as can many skills in life. Different people take different amounts of time to learn, but wanting to learn is probably the most important factor.

People often make the mistake of being either too passive or too aggressive with horses. When a would-be trainer is too passive they tend to be inconsistent with their cues, which can lead to confusion in the horse. Someone who is too aggressive can overreact to a horse's attempts to try out a response, leading to a horse that is either very nervous or that becomes dull or sour in its responses.

Good horse people share certain characteristics.

- They are consistent and give the horse clear signals, leading to the horse learning the task well.
- They have usually learned their skills from a combination of trial and error, first-hand experience and listening to others of various schools of thought.

- They are very observant and good at noticing what is going on with the horse, its immediate surrounds and in the distance.
- They can concentrate on the task in hand, which leads to clear cues and a safer situation.
- They are open-minded and willing to learn new techniques.
- They know the limitations of a particular horse and therefore how much is a reasonable amount to ask from a horse and when to stop.
- They know how to move around horses. For example, they move smoothly and positively without jerky movements. They can also adapt their movements when necessary, slowing them down for example when they are working with a nervous or inexperienced horse.
- They are assertive, not passive or aggressive. They can put aside negative feelings so that they can work a horse in the right frame of mind.
- A good horse trainer has 'feel' and 'timing'. 'Feel' describes how a trainer interprets a horse's body language signals both at a distance from it and when close. These signals are what the trainer sees and feels. 'Timing' describes how a trainer knows when to act and when not to act, when to apply pressure and when to remove it. Developing 'feel' and 'timing' happens as a trainer improves their technique.

Where to work a horse

Initially, an area that has as few distractions as possible is the safest place to work a horse. Once the horse has good basics it can be introduced to different situations. Large yards and arenas are ideal in the beginning. An open area is not. If a horse gets away from you in a larger area it can panic and bolt. A work area should have good footing and safe fencing (p. 35).

Training session length and intensity

The length and intensity of a training session depends on factors such as the age of the horse, its level of fitness, temperament, conformation and current soundness. Other factors such as the gear used, work surface and weather conditions also play an important part.

A young or old horse should not be expected to work for long periods. Also, working a horse for a long period when it is not fit or educated enough leads to stress and eventually conflict behaviour. All horses learn better if training sessions are short, followed by time to relax or by less demanding exercise.

Training horses on the ground

Groundwork includes exercises such as lunging, close contact in hand exercises, long reining, working a horse loose in a round yard etc. Most of these exercises are carried out for the purpose of training a horse. Whatever groundwork method is used, safety procedures must be followed. All groundwork exercises require the handler to have a certain level of skill. Lunging is often used as a form of exercising a horse as well for training, depending on the skill of the person doing the lunging. Some instructors also lunge a horse with a rider on board in order to improve the rider's balance and position. This should only be carried out by an experienced instructor with a well-trained horse.

Any groundwork that involves the horse being on a long rope runs the risk of the horse getting away with a long rope attached. Therefore exercises such as lunging (particularly with a rider on the horse) and long reining should be carried out in an enclosed area. Long reining calls for even more skill than lunging because two reins are used. Some horses panic and kick out at the feel of the rein around their hindquarters if they are not properly habituated (see p. 128) to the feel of the rope. Walking directly behind a horse in long reins can result in you being kicked or dragged if the horse takes off.

Any exercises involving ropes also run the risk of you getting caught up in the rope and dragged. The rope/s should never be wrapped around your hand; they should be kept up off the ground so that your legs do not get tangled in them.

Take care when working a loose horse (such as in a round yard), especially when the horse is turned as it can come too close if you don't have the skill to keep the horse to the outside of the yard.

Training under saddle

The safety considerations of training horses under saddle are the same as those for general riding (see p. 98).

Training methods

There are many methods of training horses and to the uninitiated it can be very confusing. Many trainers actually use the same methods but have different terminology that describes what they do. Two of the most useful 'tools' for training horses are 'pressure and release' and 'habituation'. These are outlined here.

Pressure and release

All horses should yield their body away from pressure when asked. In the young horse, this can start as groundwork and later on be part of the ridden training. All young horses should be taught to yield. If an older horse doesn't yield, then it should also be taught. To start with, the horse should move forward and backward, move the shoulders sideways in both directions and the hindquarters sideways in both directions. It should also lower the head when asked. When a horse will perform these movements consistently it is also possible to get the horse to stand still because you now have the tools to direct and to stop its movements. For example, if the horse will not stand still for mounting you can manoeuvre it back into place each time it moves. Eventually the horse will stand still because it has learned that it doesn't profit by moving. To teach a horse to yield, apply pressure until the horse makes the correct response and then **immediately** release the pressure. This why it is called pressure and release.

It is important not to remove the pressure until the horse makes the correct response. So, for example, if the horse steps forwards when you are applying pressure on its chest for the horse to go back, keep the pressure on until the horse actually steps back. The horse will eventually step back as long as the pressure is kept on, as the horse uses trial and error to see what it has to do to remove the pressure. As soon as the horse yields in the right direction, remove the pressure. This is how the horse learns what is required. Keep in mind that different horses respond to different levels of pressure.

Horses like to be comfortable so usually not much pressure is required to get it to react. Examples of pressure are your fingertips, taps from a whip, swinging or flicks from a rope; even your voice can be a form of pressure.

As you become familiar with teaching the horse to yield you can use your imagination to see how many different ways you can get the horse to yield to pressure.

A good place to start is to teach the horse to yield its head downwards. This is a very important part of a horse's training because it not only teaches the horse to lower the head on a cue but also relaxes the horse at the same time because the horse associates the head down position with grazing. Put pressure on the head via the leadrope or the reins, and as soon as the horse moves its head in the right direction remove the pressure. Eventually the horse should lower its head to the ground, halfway to the ground or to whatever point you ask. In the beginning don't do this on grass, as the horse will try to graze.

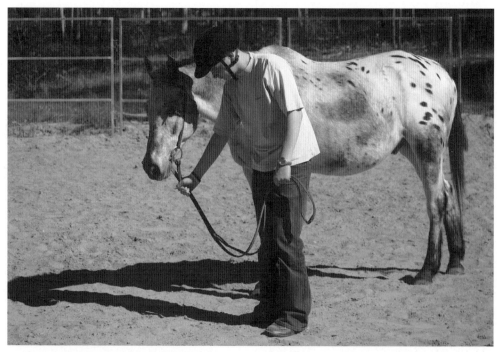

Teaching a horse to yield its head downwards

A horse that yields rather than fights is much safer both to be around and for its own sake. For example, if a horse has been taught to accept ropes around the legs and will yield the leg when asked should not panic if it gets caught up in a fence or if a rug inadvertently moves and causes the leg straps to slip around the horse's legs. A horse that has been taught thoroughly to yield its head will lead anywhere when asked, including onto a horse trailer. It will also tie up without pulling back because it will come forward (yield) instead of fighting the pressure from the halter.

Next, the horse should be taught to come forward from pressure on the halter. If you are handling the horse, one method that can be used to do this is to stand level with the horse's shoulder facing forward and ask the horse to move forward, using pressure on the halter. If the horse doesn't move forward, tap it with a whip in the girth area until it does so. As soon as the horse moves forward, stop tapping (remove the pressure). Do this over and over until the horse comes forward from just pressure on the halter. Do this in different situations to test whether the horse is yielding unconditionally. The horse should also be willing to trot forward from pressure when asked. If it will not, then keep training until it does. Teaching the horse to yield the head forward from pressure may take a while but it is time well spent because it has many benefits other than being able to tie the horse safely. The horse will lead anywhere when asked, including onto a horse trailer or truck. The more educated the horse becomes, the less pressure should be needed to get a response.

A horse can be taught on the ground to yield its mouth by you standing at the side of the horse with the rein held up towards the saddle area. If the horse walks around, keep the rein pressure on until the horse stands still and gives a soft feel on the rein (yields). Then release the rein. This should be repeated until the horse yields on both sides. This exercise can be done with a rider mounted once the horse has learned to yield on the ground.

Under saddle, a horse should yield its mouth and move forward, sideways and backwards from pressure.

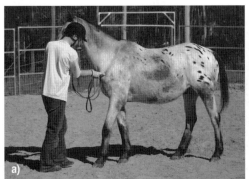

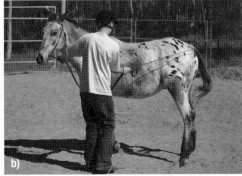

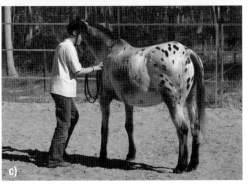

a) Teaching the horse to yield backwards.

b) Teaching the horse to yield the hindquarters over.

c) Teaching the horse to yield the shoulders over

This is just an overview on the subject of training horses to yield to pressure. The best investment you can make for both you and your horses safety is to attend a clinic run by an experienced trainer to learn the basics (see p. 118). There are also some excellent books available on this subject (see end of chapter).

Habituation

Habituation is the process used by a trainer to teach a horse to be unafraid of situations that would normally cause it to startle or behave in a way that makes it dangerous. Some people call this bomb-proofing, but it can be dangerous to think of a horse as bomb-proof. We can only train a horse to override its instincts. Those instincts are still hardwired into the brain and when a horse is frightened they may still come to the fore. The more trained the horse is the more chance there is that it won't panic, however, every horse has a breaking point.

Initially horses are afraid of most things because in the wild this is what keeps them alive. However, even a wild horse has to get used to sights and sounds that are not followed by a scary incident as otherwise it would wear itself out by using too much nervous energy.

Horses often become habituated to many common sights and sounds without the owner even realising it. For example, horses that live in an area next to a busy main road are often habituated to traffic to some extent.

You can make a horse much safer by habituating it to many different stimuli. If this is done with lots of different objects and situations the horse will also start to adopt stimulus generalisation. This means that if the horse is habituated to a whip crack, for example, it will be more relaxed if a car backfires nearby even though it is not the same thing. If the horse is habituated to being touched by a plastic shopping bag it will be calmer about other flapping objects in the future. Remember, though, that what we see as similar objects may not be seen the same way by a horse.

Habituation will make the horse much safer but it is important that it is carried out properly, otherwise it is possible to actually make a horse more rather than less frightened of scary things. If a horse is allowed to run away from something scary it is learning a very dangerous behaviour. Therefore, before a horse is introduced to new sights, sounds and feelings it must be taught to stand still. If you are the handler, you must have control of the horse's movements because if you don't and a scary situation occurs the horse can run away. The horse will have learned to be afraid of the situation. See the previous section on yielding, and make sure the horse is proficient and that you can make the horse stand still before beginning the process of introducing scary things. Introduce new objects in an enclosed area such as a training yard or fenced arena.

Mounted police use habituation extensively to get horses accustomed to the many situations that they have to face as part of their job. See if you can get a group together and go to a training session at your local mounted police. Alternatively, it may be possible to get a retired mounted police officer to come and give a training session to a group.

The horse should be habituated to both external stimuli and to the touch of objects that may at first scare it. Therefore the horse should eventually accept the *sight* of things such as:

- pram/pushers, still and moving;
- umbrellas, being put up and down;
- flapping tarps and flags;
- plastic bags blowing around;
- dogs and other animals;
- traffic of all types;
- children playing and children's toys.

It should also accept the *sound* of such things as:

- motors, so that it is not frightened of the sound of traffic;
- load noises, sharp bangs and cracks (whip cracks), because these can occur anywhere and anytime;
- hissing sounds such as aerosols;
- crackling sounds such as tarps and sheets of plastic.

It should accept the *feel* of such things as:

- being touched anywhere, at first just with the hand, later with a saddle cloth, tarp or plastic, straps and ropes, water etc.

Most of these situations can be set up at home, therefore use your imagination to add to this list of objects:

- a feed sack, a tarp and a plastic sheet – eventually a horse should accept these anywhere on the body and should also walk over them;
- a long rope – this can be used to habituate the horse to the feel of straps around the legs (such as when a rug slips) and to teach the horse to lead by the leg, which reduces the chance of it panicking if caught up in a fence;
- streamers hanging from a branch of a tree and a plastic bag tied to a whip or stick – these teach the horse to be unafraid of flapping objects;
- tyres to step over and through – these teach the horse to put its feet in a dark space;
- a stock whip for cracking – this teaches the horse to not panic at a loud bang or crack;

Rubbing a horse with a feed sack on the body – a new situation.

Work up to the head once the horse accepts the sack on other areas

- water on the body and the legs – this helps the horse to accept water in different situations;
- a feed sack filled with aluminium drink cans being rattled;
- a Pilates ball/beach ball.

The more potentially scary things that a horse is accustomed to the safer it becomes.

To habituate a horse to new sights, sounds or feelings, start by introducing at a low level of pressure and only step up the pressure when the horse has accepted the previous level. Initially this may require standing at a distance from the horse while it is secured in a yard or stable. Never frighten the horse. Introduce the new objects at a level that the horse can cope with. It is vital to remember that if the horse is allowed to run away you have taught the exact opposite of what you wanted and made the horse less rather than more safe.

Recommended further reading
There are some excellent books on the subject of training horses. The following books all deal with the subject of yielding to pressure and habituation to scary objects. The authors of these books (with the exception of Tom Roberts) also run clinics, so look out for them.

Brady S (2005). *Horse training: Steve Brady's formula for success*. Steve and Linda Brady, NSW, Australia

Chatterton J (2000). *John Chatterton's ten commandments*. John & Janet Chatterton, Queensland, Australia.

McLean A & McLean M (2002). *Horse training the McLean way: the science behind the art*. Australian Equine Behaviour Centre, Victoria, Australia.

Roberts T (1988). *Horse control series* (four books). Greenhouse Publications, Melbourne, Australia.

9

Transport

Transporting horses is potentially a very dangerous situation. This chapter looks at the safe transportation of horses. The safety of horse handlers and drivers is considered paramount, so the chapter focuses on issues that will keep people safe when dealing with horses in such situations.

From the handler/driver point of view, transporting horses can be stressful if they are not confident about loading and driving a truck or towing vehicle containing horses. One bad experience can make the whole situation worse in the future for both the people and horses concerned. From the horse's point of view, the business of transporting offers many reasons to be fearful and possibly panic. Horses naturally avoid confined spaces and unstable surfaces, both of which occur when transporting. Unfamiliar noises and smells add to the frightening experience.

Good preparation of the vehicle and equipment, correct training of the horse, practice sessions and increased handler/driver experience all vastly decrease the chances of things going wrong.

If you are inexperienced, get help from a professional or experienced horse person, especially if having problems such as loading a difficult horse.

Vehicle safety

Horses can be transported in a truck that is adapted for horses or in a horse trailer towed behind a suitable towing vehicle. Trucks are usually partitioned so that the horses travel sideways or on a slight angle. Trailers are partitioned so that the horses face forward or stand on an angle (angle load trailers).

General considerations

Trucks, towing vehicles and horse trailers must be roadworthy and road legal. A mechanic should regularly check trucks, towing vehicles and horse trailers for mechanical defects, including the chassis.

Trucks and towing vehicles

The mirrors on a truck or tow vehicle should be large enough so that the driver can see well back behind the body or the trailer. Extensions can be fitted to existing mirrors if necessary.

When towing a horse trailer, be aware of the maximum legal weight that the towing vehicle can tow. The vehicle towbar must have the correct rating for the weight of the laden horse trailer (this should be stamped on the towbar). Heavy-duty towbars are required for towing horse trailers. Never use poor-quality and/or homemade towbars. Check the tightness of the bolts and studs that attach the towbar to the vehicle at least twice a year. Bars must also be fitted with attachments for safety chains.

Visual checks

An owner or person responsible for the truck or horse trailer should visually check inside and outside the body of the truck or trailer before and after each journey. Checks should involve looking for any signs of wear and tear such as rust, welding coming apart, the floor rotting, lifting metal, rivets coming undone, damaged door hinges or fasteners and cracks in the hitch (with a trailer) and tyres. The area where the horses stand should be thoroughly cleaned out after every use, sweeping or hosing out as required.

Projections

Inside and outside the truck or horse trailer there should be no projections that could injure handlers or horses. Bolts and fixtures should be flush with the wall. Remember that horses can get into unusual positions when travelling – any projection is potentially

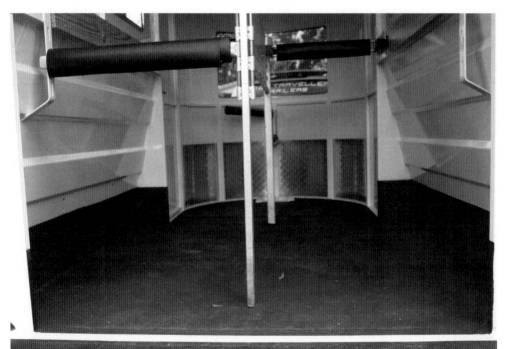

A rubber-covered floor in a trailer

dangerous. In particular, sharp edges on the protruding mudguards of horse trailers can cause injuries to handlers or horses.

Flooring

The flooring should be constructed of a strong material and be checked frequently for rotting or corrosion. The place where most floors show wear is at the edges where the boards meet the walls. Wooden floors are worn much more quickly if horses wear shoes that are studded or that have the heels turned down. Rough-surfaced heavy-duty rubber is ideal as a floor surface. It can fixed and sealed so that it is easily hosed out after use. If it is removable, it needs to be taken up and the floor cleaned every time the truck or trailer is used to prevent manure and urine from rotting/corroding the floor.

Tyres

The tyres of trucks, towing vehicle and trailers should be in good condition and have the correct tyre pressure (check manufacturer's recommendations). Light truck tyres are usually safer than car tyres on horse trailers as they are more robust. One spare wheel is essential, but having two is safer for trucks and trailers. They must be kept in an accessible location and in good condition.

Partitions

Partitions that separate horses should **not** go right to the floor because that prevents horses from spreading their legs to balance. A horse needs to spread its legs further out than the width of its body in order to balance when cornering. This is difficult to achieve

A shoulder partition extends back from the centre pole to prevent horses from getting their head on the wrong side while travelling

in forward-facing trailers as the outside pair of legs are up against the outside wall of the trailer. The driver must be extra careful when cornering with a forward-facing trailer as the horses can panic (scramble) if they can't spread their legs out to balance.

Other partitions include head dividers (stallion dividers) that prevent horses from biting each other while travelling. In forward-facing horse trailers a partition that extends backwards from the centre pole (shoulder partition) is a good addition as it prevents a horse getting its head stuck on the wrong side of the centre pole while travelling. Getting stuck can happen even when a horse is correctly tied and it is very dangerous for the horse.

It is useful if partitions are removable but they should **not** be able to be lifted off their fixings when a horse leans or rubs its head on them.

Internal lights

Internal lights make it safer to load horses in the dark. If horses are loaded in the dark, remember to give their eyes time to adjust to the light before asking them to step into the truck or trailer. Internal lights should be situated where horses cannot knock them with the head.

Dimensions

The internal physical dimensions of any horse transporter should be large enough to accommodate the horse/s. Full-size horses (16hh plus) need the roof height to be a minimum of 7 foot (2.10 m). Ponies and smaller horses can manage with less roof height (around 2 m). The stall length should allow the horse to stand comfortably and to stretch the neck forward. A common complaint about many angle-load trailers is that the stalls are too short for larger horses.

Safety aspects of horse trailers

Horse trailers can be forward-facing or angle-load. A horse trailer should have safety chains welded onto the drawbar. The chains are attached to the towing vehicle's towbar with shackles when it is hitched. They help to keep the horse trailer attached to the towing vehicle in the event of it detaching from the towball when travelling.

Horse trailers should have clearance lights high on the front plus reflectors, stop, tail, indicator and numberplate lights on the rear. Check with the relevant transport department.

Horse trailers must have a jockey wheel. The jockey wheel makes the horse trailer stable when it is not hitched to the towing vehicle. It also helps when hitching the horse trailer, by enabling the tow hitch to lower onto the towball without you having to lift it. It is removed or held up out of the way with a swing-up bracket when towing the horse trailer. A swing-up bracket is also useful so that the jockey wheel does not get lost when not in use. Never attempt to lift a trailer onto or off a towball without the use of a jockey wheel as it can cause a back injury.

A horse trailer should have double axles (and therefore have four wheels rather than two). Single-axle horse trailers are dangerous (and illegal in many countries) because if a tyre blows out during travel the horse trailer is likely to flip over. A dual-axle horse trailer is more likely to stay stable until the driver can stop safely.

A horse trailer should have one or more doors at the front of the trailer for the entry and exit of the handler when the ramp is closed. These doors are sometimes called

A swing-up bracket for a jockey wheel

grooms' doors. They should open outwards and preferably be full height so that the handler does not have to bend down when entering or leaving the trailer. A truck does not always have grooms' doors leading to the outside, however, there may be access to the horse section from the living area or cabin.

Trailer brakes

Horse trailers are fitted with either electric or hydraulic brakes. Brakes should be well maintained and the correct type for the weight of the trailer. Horse trailers with brakes on all four wheels rather than two are safer; this is usually a requirement on heavier horse trailers. Certain size horse trailers may be required to have an additional breakaway system that stops the horse trailer and holds it for at least 15 minutes in the event of it detaching from the towing vehicle. These are a good safety measure on all trailers with electric brakes. If the trailer does have electric breakaway brakes, there will be a battery on or in the trailer. This needs to be checked frequently. Check with the relevant department of transport about regulations for horse trailer brakes.

Horse trailer brakes should be adjusted so that they slow the horse trailer simultaneously with the towing vehicle brakes. This will let the vehicle and trailer pull up as one unit, not as one part pushing or pulling the other to a stop. Electric brakes may need to be adjusted for an empty trailer, for one horse, for two horses etc. There is usually a device for adjusting electric brakes near the drivers' seat in a vehicle that has been equipped to tow a trailer with electric brakes.

Above the ramp

A truck ramp usually covers the entire height of the opening. Horse trailer ramps only partially cover the back entrance to a trailer. A cover above the ramp of a horse trailer reduces noise (which can frighten a horse) and can prevent a horse from jumping out if it

A lift-up back door on gas strats.

Barn doors on an angle load three-horse trailer

A trailer with adjustable and removable breast and rump bars

breaks free. When transporting foals, it is essential that the area above the ramp be enclosed because a foal can turn in a small space and may try to escape over the ramp (even when its dam is with it). The cover can be a tarp that rolls up when not in use or some form of solid doors.

A forward-facing horse trailer has breastbars that prevent the horse from falling forward. Removable breastbars are safer than fixed breastbars because they can be usually be removed if a horse becomes trapped under or over them (depending on how they attach). The breastbar should not lift off its fixings by itself if a horse gets underneath it. Breastbars should be the correct height for the horse. A common problem is breastbars that are too high for ponies, that can then get trapped under them. Therefore breastbars that adjust up and down are safer.

In forward-facing horse trailers there should be safety chains, bars or breeching doors to contain the horses within the trailer after they have been loaded and before the ramp goes up. These also prevent the horses from stepping backwards before the handler is ready, after the ramp is lowered to unload the horses. Chains and bars need to be height-adjustable if you are planning to transport different sizes of horses. Small ponies can back out underneath if the chains and bars are too high. If they're too low, larger horses can flip backwards over the chain or bar if they start to move backwards after the ramp is lowered for unloading. Breeching doors are deeper than a bar or chain, which means that they fit a larger range of horses and ponies without adjustment.

Ramps

The ramp of a horse truck or horse trailer should have a non-slip surface. Between the top of the ramp and the vehicle, there should be no gaps that a handler or horse can trap a foot in. If there is a gap it should be covered with a secure strip that will prevent feet or hooves from getting stuck. The ramp on a truck tends to be steeper than on a trailer because the body of the truck is much higher off the ground. Ramps that are steep can be fitted with steel props underneath to make them

A non-slip ramp

less steep. Some trailers have a short steep ramp which tilts a horse too far upwards when it places its front feet on the ramp. This gives the horse the impression that it will bang its head on the roof as it goes in. This is often the beginning of a horse becoming a problem loader. Again, steel props can improve this type of ramp.

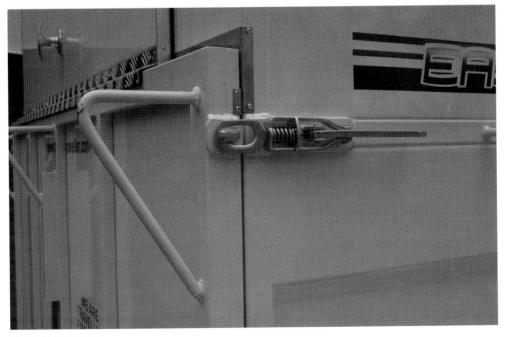

Steel props on a ramp

The ramp needs to be spring-loaded or have a mechanical system for lifting and lowering. A ramp that does not have any lifting and lowering assistance is dangerous because it can cause back injuries, or drop too quickly and heavily injure anyone in the way. Always use correct lifting procedures when operating a ramp (see p. 17).

Windows

The front window of a forward-facing horse trailer should not be large, however, many are supplied with a large window. When travelling on narrow roads the oncoming traffic can suddenly appear too close to the head of the horse on that side. If a horse starts to panic (for whatever reason) and tries to get out of the horse trailer, the large window is an invitation to it. A large window can be improved by welding a grille on the inside or by taping strips of duct tape across and down to give the illusion that there is a barrier.

Side windows placed high on the side walls of horse trucks or horse trailers are useful for ventilation and natural light. Good ventilation is very important for reducing the incidence of travel sickness in horses. Horses generate a lot of heat, especially when travelling, as they can get stressed and on longer journeys fumes from manure and urine can build up inside the transporter. Vents fitted in the walls or roof can offer additional ventilation. Make sure that the ventilation system does not funnel vehicle exhaust gases to the horses. You can buy an extension for the vehicle's exhaust pipe that channels the fumes away.

Hitching a trailer

Hitching a horse trailer involves a series of steps that are carried out in reverse when unhitching. First, back the vehicle until the tow hitch is over the towball. It helps if someone guides the driver by standing where they can see the hitch and the driver can see them. Wind down the jockey wheel until the hitch is coupled. Then release the lock that secures the hitch to the towball. It is easy to forget to do this and drive away without the lock, which can result in the horse trailer becoming unhitched in transit. Then wind the jockey wheel all the way up and remove it. The lights socket can then be plugged in and the safety chains attached. These chains should be long enough to cross each other under the drawbar so that if the trailer comes off the towball they support the trailer and stop it from nose-diving into the road. Then remove the handbrake on the horse trailer. Check light globes and wire connections after hitching, each time the horse trailer is attached. The indicators, reversing lights and brake lights must all be working before beginning the journey.

The hitch heights of the towing vehicle and the horse trailer must be matched to ensure the safe handling of the vehicle and the comfort of the horse. This should be checked with the horses on board, although in some cases it is apparent before the horses are loaded that

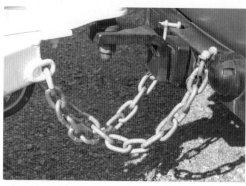
Safety chains

(a) Hitch is too low.

(b) Hitch is too high.

the heights will be unlevel. The horse trailer should not pull or push the back of the towing vehicle up or down. In both cases the steering and suspension of the towing vehicle will be affected, the tyres receive uneven wear, the headlights of the towing vehicle will be directed at the wrong angle and the horses forced to stand at an uncomfortable angle in the trailer.

To solve this problem, check that the towing vehicle is the right size in terms of weight and power, and that the manufacturer's tow rating is not exceeded. If the towing vehicle is too high an adjustable-height towball can be fitted so that the horse trailer is not pulled upwards at the front. A level-ride device, known as a weight-distributing hitch, will make sure that weight is transferred in the right places across both vehicle and horse trailer. It can be purchased and fitted at towbar specialist shops. Other forms of levelling devices include overload springs and equaliser bars that can be purchased from specialist shops and fitted by a mechanic.

A Hayman Reese weight-distributing hitch

Unhitching a trailer

Unhitch in a flat area. To unhitch a horse trailer apply the horse trailer handbrake, undo the safety chains, unplug the lights socket, release the towball lock then attach the jockey wheel and wind it down until the tow bracket lifts off the towball. Hang the lights plug in such a way that it will not get wet in the rain and corrode.

Packing for a trip

Pack in advance so that you can leave without delay once the horses are loaded. Some horses tend to fidget when they are loaded and their paddock companions may be calling to them, causing them to get agitated also. Have a full tank of fuel before loading the horses and check tyre pressure at the same time. Tyre manufacturers base their recommended inflation pressures on tyre size and load-carrying capability, and these are available from tyre dealers. Check the tyres often for irregular wear, bulges, cuts, nails and so on.

Whether you need to pack feed and water, grooming equipment etc. depends on the length of the journey and the destination, however, certain safety items are necessary whatever the length of journey. Items that should be packed for every journey include:

- separate first aid kits for humans and horses;
- a pocket knife for emergencies;
- a small toolkit, wheel brace, suitable jack and a sloping block of wood (see p. 149);
- a mobile phone – but remember that there is often poor or no reception in country areas;
- a spare halter, leadrope and twine;
- a flashlight even if you're not planning to be out after dark;
- a fire extinguisher;
- spare fuel;
- a reflective triangle and reflective cones in case of breakdown.

Horse travelling gear

A horse can be fitted with various items of gear to protect it from injury during the journey. Areas that tend to get injured are the legs, tail and poll. These can be protected with items such as boots, tail/leg bandages, poll guard etc. Horses that are travelling with other horses standing beside them have a much greater chance of injury to the legs.

When fitting travelling gear, secure the horse safely or have a competent person hold the horse. Never kneel on the ground beside the horse when fitting or removing boots or bandages – bend or crouch instead. This enables you to get out of the way if the horse moves suddenly. Stand to the side of the horse (rather than behind it) if fitting anything to the tail. Always use a strong halter and a leadrope that has plenty of length (see p. 41).

Loading horses

Loading (and unloading) horses is a situation where handlers should consider wearing a helmet even if they don't normally wear one when handling horses. Handlers should also wear strong boots with grip on the sole. Gloves may also be necessary. Because of the potential dangers of transporting horses, it must be emphasised how important it is that **every** horse learn to load and travel safely not just for its owner but for any reasonably experienced person. Even if a horse is not required to travel regularly to compete it may have to be transported occasionally, for example to escape a fire or flood or to go to a veterinary surgery for emergency treatment.

With any horse the aim is that it should load with the minimum of fuss. Ideally the horse should be able to be loaded by one person, however, it is safer to have at least one other person nearby in case anything goes wrong. If you are the handler, bear in mind when loading a horse that is new to you that some horses will load for a familiar person but not for an unfamiliar one. Two different people don't give identical cues, which could be enough to confuse the horse and lead to hesitation in loading, until the horse begins to understand the new cues and responds accordingly. Some horses will walk willingly into a familiar truck or horse trailer but are suspicious of a strange one. What seems like minor differences to us are not minor to the horse.

A horse must have well-established basics in order to load. The basic requirements are that the horse moves forward, backwards and sideways from pressure **whenever** and **wherever** asked (cued) and that the horse will stand still when asked (cued). Many horses do not and this is when problems occur with loading (see p. 144).

One of the most common situations requiring transport involves taking a horse to a show or event, but this is not an ideal learning situation for an inexperienced handler or horse. Trying to load a horse to get somewhere on time, or at the end of a long day at a show or event, is very stressful as the handler may not be in the right state of mind. Frustration, tiredness or disappointment are common emotions at the end of such a day and even the euphoria of a good day can soon evaporate if the horse will not load. Even if you manage to load the horse eventually because other people lend a hand, the horse might have learned the wrong behaviour in the process.

If you are an inexperienced handler, get a more experienced person to help you practise loading well before the outing in question. Don't set off on an outing if you're at all unsure about loading the horse at the end of the day. Have lots of driving and loading practice sessions before the outing. For example, drive to a friend's property with the horse and practise reloading in a different but less stressful environment than a showground.

A loading area should be flat and clear of obstacles. The loading area needs to be secure (in case the horse gets away) and any unnecessary people and loose dogs should not be there. Don't park next to a fence or wall unless they are solid and high (at least 2 m). A horse can topple over a low fence if it jumps off the ramp when loading. Out in the open (away from obstacles) is safer if the horse has the basics established, such as leading forward, backwards and sideways. A horse should load without the use of barricades etc. Remember, never load a horse into an unhitched trailer because it is unstable and will move.

Loading into a truck or angle-load trailer

To load a horse into a truck or angle-load trailer, lead the horse towards the ramp with you walking between the head and the shoulder. If the horse wants to stop at the base of the ramp to sniff it, allow it to do this. Do not start tugging at the horse's head. Then ask the horse to walk up the ramp by giving the usual walk forward cue. Once inside, put the leadrope over the neck of the horse and, still holding the horse with one hand, manoeuvre the horse around so that the partition blocks it in. The partition must be secured **before** the head is tied.

Loading into a forward-facing trailer

The front (grooms') door can be opened to let light inside but there must be a barrier (breastbar) to prevent the horse from trying to get out the door as it loads. Again, lead the horse towards the ramp and allow to sniff it if it wants to. Then cue the horse to walk forward.

A well-trained horse should walk past you and into the trailer as you put the leadrope over the horse's neck. The horse can then be secured behind with the chain, breeching bar or breeching door before you go to the front of the horse to tie the head.

If the horse will not load without you going in first, walk up the left-hand stall as the horse loads into the right. Do not duck under the breastbar. Aim to train the horses in the future to walk into a trailer alone.

In all cases a horse must be secured behind before the head is tied. This is because if the horse pulls back and there is nothing secure behind it, the horse will pull until

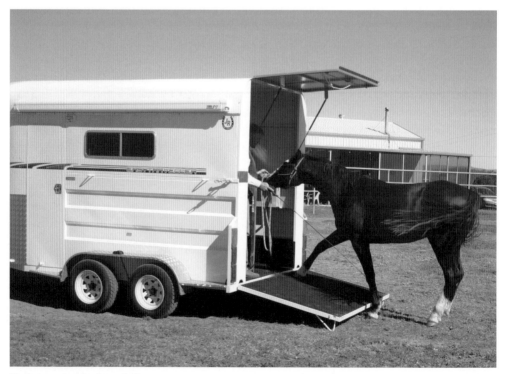

The horse enters the horse trailer

something breaks, usually banging its head on the roof at the same time. Once free the horse will run backwards down the ramp and may injure anyone in the way.

When securing a horse with a chain, rump bar or breeching door always stand to the side rather than directly behind the horse. This will allow room to move out of the way if the horse suddenly runs backwards.

If you are holding a horse in a transporter while waiting for someone to fasten the horse in behind for example, never pull the horse forward if it starts to pull back. The more the horse is pulled the more it may pull back, which can result in serious injury for the handler.

Many people tie the leadrope to a loop of twine when tying the horse inside the truck or trailer. The problem with this method in a trailer that does not have doors above the ramp is that if a horse breaks the twine and gets free it could try to escape over the ramp. The horse should be tied using a quick release knot that can be undone quickly in the case of an emergency. You should always have a pocket knife handy just in case. It is important that the horses be tied short enough that they cannot spar with each other or get their heads around the wrong side of the middle divider pole (in a trailer). At the same time they need to be tied long enough that they can lower the head to clear their airway, especially on long journeys. This usually means about 50 cm from the tie ring to the halter ring, however, it will vary depending on how high the tie rings are fitted.

If you are loading two horses it can be a good idea to load the most sensible or experienced horse first. However, this will mean loading the second horse without being able to walk into the trailer with it. Remember, do not duck under the breastbar. Often an inexperienced horse can be coaxed on by the presence of another horse. On occasions, however, this will discourage an inexperienced horse from loading because you are asking it to walk into a narrow space next to a horse it may not know. Without the other horse in place the centre partition can be moved over to create more space. Decide what is likely to work best on the day.

If transporting two horses in a forward-facing horse trailer, put the heavier horse on the side nearest the middle of the road to even out the effects of the camber (the slope of the road from the middle to the sides) and give better balance to the horse trailer when travelling. Likewise, if transporting a single horse put it in the stall nearest the middle of the road.

Remember the following rules.

- Practice and training in the basics is vital for successful loading.
- Always secure the horse behind **before** tying the head on loading and release the head **before** unfastening the chain, rump bar or breeching door or partition (in a truck) or angle load trailers when unloading.
- Always stand to the side rather than directly behind the horse when securing and unfastening the chain, rump bar or breeching door.
- Always stand at the side of a ramp rather than at the base to lift it and lower it. Use correct lifting techniques (see p. 17).
- Never tug on the head of the horse when trying to load it or once it has loaded, as this can cause it to rear and hit its head, or run out backwards.

Training a horse to load

Aim to minimise the time you have to stay inside a trailer or truck with a horse by training it to load quickly and quietly and preferably without you having to walk inside with it. Training an inexperienced horse to load is far easier than retraining a horse that has learned the wrong behaviour. In order to learn to load, a horse must first have very well-established basics which includes moving forward, backwards and sideways from pressure. This pressure should include the halter, a hand and light taps with a long thin stick or whip. For example, a horse should move forward from pressure on the leadrope and from taps with the whip to the girth area. It should move the shoulders away from taps with the whip to the top of the front leg. It should move backwards from several cues such as a push with your hand on its chest or nose, taps on the chest with a whip or even vibrations with a leadrope. It should move the hindquarters over from taps with a whip to the top of a hind leg, a push with your hand etc. The more cues a horse will respond to the more 'tools' the handler will have.

Practise these cues over and over until the horse responds automatically (see p. 125). Before asking the horse to load into a transporter, practise the responses to cues by asking the horse to move forward, backwards and sideways in a variety of situations. Any gaps in this basic training will show up as soon as the horse is asked to do something that it does not want to do, such as load. The reason that so many horses have a problem with loading is because they do not have the basics established before being asked to load.

It is also useful to ensure that the horse has been in an enclosed area before loading into a trailer or truck as the flight instinct is strong and the idea of being confined often scares a horse that has spent its whole life in the open. Time spent in a stable or even under an overhanging veranda assists in familiarising it to being enclosed and confined.

Always use a good strong halter and leadrope. A long thin stick or whip is also needed to lightly tap the horse with.

Once the horse has well-established basics and will respond to cues in various situations, the truck or horse trailer can be introduced. Do this on a day when you aren't planning to travel anywhere and treat it only as a training session. Ask the horse to step up the ramp after allowing the horse to have a look and sniff at it. If the horse will not lead forward, tap with the whip in the girth area until it takes one step forward. As soon as the horse steps in the right direction **immediately** stop tapping. If the horse goes backwards at any time, tap until it stops (this will mean walking backwards with the horse). If the horse moves the hindquarters sideways tap them straight again. Never tug on the head. Be patient. Never allow the horse to turn around and walk away from the ramp and keep asking for small steps in the right direction. Always reward any tries immediately by removing the pressure (taps) and allowing the horse to relax before asking again.

This method utilises 'pressure and release'. Pressure is applied, in this case small taps, and removed when the horse steps in the right direction. The most common mistake that people make is that they do not remove the pressure quickly enough when the horse moves in the right direction. They keep asking for more without letting up on the pressure – because the horse is not being rewarded for steps in the right direction it does not learn the correct behaviour.

Be patient and quietly persist, and the horse will eventually load. Practise loading and unloading several times before introducing travelling. With practice you will be able to control the horse's movements so that it can be asked to walk forward and backwards and stop at any point on the ramp or inside the transporter.

The method described above is only one way to train a horse to load. There are other methods that work just as well, used by many respected trainers, and all these methods are equally valid. They all work on the basic concept of pressure and release. It is up to you to decide which method is preferable for you and your horses.

Horses that are difficult to load

There are numerous reasons why a horse can be difficult to load.

- The horse has got away with refusing to load in the past and has therefore learned not to load.
- The horse has had bad experiences with loading and travelling and is now frightened.
- The horse and/or handler is inexperienced.

Make sure the horse has no reason to be afraid to load. The transporter should be safe, spacious and solid. If it is not, then you can't blame the horse for refusing to get in it.

Horses that will not load can be taught to load in the same way that an inexperienced horse is taught. The solution is going back to the basics outlined above and teaching the horse to respond unconditionally to these cues. Only then should loading be introduced. At this stage the horse will still probably resist loading if this behaviour is entrenched but it will eventually respond to the cues if they have been taught thoroughly. Be patient, because this type of horse will be more difficult than an inexperienced horse that does not have established avoidance behaviours. Be careful in this situation – horses that have learned bad habits can be very dangerous and react violently. If you are at risk, call on someone more experienced at managing this type of situation.

There are occasions when a horse may have to be loaded without time to train it beforehand, for example at a saleyard. It is possible to still load a horse successfully in this situation but bear in mind that the horse is not learning to load and will need to be trained later when the situation allows. These methods require the help of other people and equipment.

The lunge line method requires the help of two experienced and confident people (wearing non-slip boots, gloves and riding helmets) and two lunge lines. Fasten the lunge lines to either side of the ramp (the ramp fasteners can be used), and cross them behind the horse with a person holding the end of each line, standing well forward (not behind the horse). Lead the horse forward as the assistants put pressure on each line so that so that the horse is coaxed into the truck or horse trailer. This method is usually very effective as the horse feels the pressure from the lines and usually goes forward.

If previously unhandled horses have to be transported they should travel loose in a truck or large trailer, like cattle. Do not attempt to barricade or tie up horses with no previous experience of such. They can be loaded via yards or a cattle ramp if necessary.

Never resort to sedation or blindfolding to load a horse. Sedating horses for transport is not a good idea and should only ever be done by a vet in special circumstances. Blindfolding is very dangerous for the handlers and the horse, because a horse can crash about and even fall when wearing a blindfold. Horses have been known to fall sideways off a ramp and kill people when blindfolded.

Unloading

As for loading, a safe unloading area should be not slippery, it should be enclosed in case the horse gets away and there should be no bystanders who could get in the way or get injured.

Before lowering the ramp, make sure that the horses are upright and still secure. With a truck or a trailer that has a tarp or doors above the ramp this will involve checking them from the front (via the grooms' door). Otherwise it may be possible to check the horses by carefully looking over the ramp. With a forward-facing trailer check that the horses are still secured behind in case, for example, the chains have come undone while travelling.

With a forward-facing trailer, lower the ramp and take care to stand at the side rather than directly behind the horse. Then go in the front (grooms') door, untie the head of the horse and put the leadrope over its neck before going around to the back again and unfastening the chain, breeching door or rump bar. Never stand directly behind the horse; always stand to the side to do this. Be careful when undoing a bar or breeching door as it will still be fastened to one side and can swing back and hit you if the horse is leaning on it or suddenly jumps backwards. If the horse is leaning on the bar or breeching

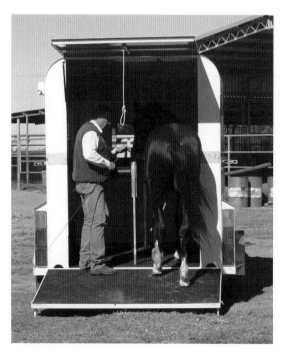

Unloading the horse from the trailer

door, cue it to move forward before unfastening. A horse should be trained to stand and wait for a command to back out rather than backing out as soon as it is released behind. Give the horse the cue to back out of the trailer. To prevent the horse from slipping off the side of the ramp, stand at the side of the ramp and guide the horse by putting a hand on its hip. As the front end of the horse emerges take hold of the leadrope that is over the horse's neck.

In the case of most trucks or an angle-load trailer the horses are positioned side-on or on an angle. Undo the horse's head **before** releasing the partition. The horse can then be led down the ramp.

When unloading more than one horse, do not lead the first horse out of sight of the horse left inside as that horse may become anxious. Keep the first horse in sight until the other horse is unloaded.

Problems with unloading

Training a horse to unload safely is just as important as training to load. Many people do not spend enough time on this aspect of travelling.

The most common problem when unloading is that some horses rush backwards out of a horse trailer. This is dangerous because they can run over a person standing in the area. Some horses, if not secured behind properly, will rush backwards before the ramp is fully lowered. This situation is even more dangerous as the ramp can hit a person, even crushing them underneath. This problem occurs through incorrect handling. A horse that is accidentally left tied at the head when released behind will start to back out but will panic when it feels the pull of the rope. This leads to the horse pulling back and maybe even banging its head on the roof. The next time the horse is unloaded it rushes back because of the scary experience last time. The habit becomes ingrained if the horse rushes each time it is unloaded.

The solution lies in training the horse in the same cues that teach the horse to be a good loader (see above). The horse must be taught to walk forward and backwards at any point inside or on the ramp of the horse trailer. With a horse that rushes backwards, you will need to spend extra time training it to walk part way up the ramp and then back off. Progress until it is possible to get the horse to back out of the horse trailer and stop part way down the ramp, using taps of the whip to signal when the horse should stop going back and should come forward.

Travelling safely

Driver and passenger safety, the safety of other road users and the horses' safety depend on careful and competent driving. Towing a horse trailer or driving a truck is much more stressful than ordinary driving and therefore requires more knowledge and skill.

Do not attempt to transport horses unless there is an experienced driver. The driver needs to be able to drive competently and confidently before taking on the extra responsibility of driving a truck or towing a horse trailer. Inexperienced drivers must always have someone experienced with them when transporting until they are competent and confident. Even when you are experienced, it is always safer to have at least one companion, especially if you are transporting more than one horse.

One way to gain experience at transporting horses is to go with an experienced driver who is transporting horses, in order to learn from their experience. The next stage is to practise driving with an empty truck or towing an empty horse trailer to gain experience in driving a larger vehicle and/or towing. Practise the necessary manoeuvres such as reversing. Then progress to short trips with horses on board.

Driving techniques

Poor driving technique is a major cause of accidents when transporting horses. Smooth driving is therefore essential when transporting horses.

- Don't let other road users pressure you into driving faster than you are comfortable with or even at the speed limit if you don't feel it is safe to do so. Drive at a speed that takes into account road surface, side winds, the weather, traffic conditions and your abilities.
- Never weave between lanes as this will swing the horses around inside the transporter.
- Acceleration and braking should be smooth and gradual to minimise excessive forward and backward movements of the horse. Never brake hard unless it is an emergency.
- Change down to a low gear when travelling downhill to slow the vehicle without excessive braking. When travelling uphill change down before the engine starts to labour.
- Remember to allow plenty of braking distance from vehicles in front and to look even further ahead than normal. Be aware that the extra space in front of you is an open invitation for other road users to jump in, so allow for this as well. When stopping or overtaking, remember the extra length and weight of the vehicle.
- If there is any doubt about overtaking, don't! Allow any queued traffic caught behind you to pass by pulling over when it is safe to do so. Be aware, however, that it may be difficult to re-enter the traffic flow.
- Poor cornering technique results in the horse falling over or 'scrambling' as it struggles to stay upright. This soon becomes a habit and before long the horse will scramble every time it anticipates a corner. Poor cornering technique is exacerbated for horses travelling in horse trailers because trailers move more than ridged truck bodies around corners. Slow down before a corner, keep your speed low and don't accelerate at all while going around the corner, then check the mirrors to see that the horse trailer has fully straightened out before picking up speed again. When cornering, allow for the fact that the horse trailer 'cuts in' on corners and curves by taking them wider than usual.
- When driving (a vehicle and trailer), bear in mind that the shock absorbers of the horse trailer are probably not as good as the ones on the vehicle and that the horse may be having a far less comfortable ride than you. Bumps in the road, gusty winds, excessive speed and road camber contribute to sideways movement, which a horse (being longer from nose to tail than from side to side) will find difficult to counter.
- When approaching traffic lights, keep your speed low and if the lights change to red roll slowly towards the lights. If the lights change to green again, you will avoid actually stopping and can have slow and smooth acceleration through the lights. This is useful when transporting horses that start to fidget and kick as soon as they feel the horse transporter stop. You may also avoid a standing start on a hill.
- When reversing, be aware that the trailer has large blind spots. Try to find someone to watch you if possible. Also be aware of anyone between the trailer and the vehicle, perhaps checking the hitching for example.

Driver comfort and safety

Planning the route is a vital part of transporting safely. Plan the journey so that you are not rushing to be somewhere on time. Frequent stops are vital for driver concentration.

If possible, travel with a companion or as part of a convoy if it can be arranged. On longer trips it is very useful if the companion is experienced enough to share the driving.

Be prepared, as the journey will take longer than if driving an ordinary car and remember that more fuel will be used. Make sure that you know the route from start to finish. This will reduce the number of 'indecision' stops as well as ensuring that you don't get lost and travel extra kilometres. Plan travelling times to avoid peak hour traffic congestion and stop-start-wait traffic. Plan the route to include straight roads where possible. Even if this adds distance, the driver and the horses will be less stressed.

Horses that travel badly

Some horses 'scramble' when travelling, attempting to balance around corners. If the horse cannot spread its legs out further than its body it can panic and start to struggle in its stall.

A horse trailer should not have full-length (to the floor) partitions so that the horse can spread its legs towards the inside. However, the horse will not be able to spread its legs on the side that is against the outside wall and some horses tend to scramble when cornering to that side. Solving this problem can sometimes be as easy as taking out the

partition in a double-horse trailer, which gives the horse room to move away from the wall and spread all four legs. If two horses are traveling together, the centre partition can be removed, as long as the horses are well acquainted. Many horses travel better this way. Removing the partition usually means that the breaching doors etc also have to be removed, so only do this with horses that have been trained to stand still inside the trailer. Some horses will scramble when on one side of the trailer but not on the other, so experiment with placing the

A JR Easy Traveller trailer

horse on different sides. Scramblers will usually travel well in a truck or angle-load horse trailer because they can then spread their legs out on both sides. There is a special patented horse trailer that has flared outside walls. This allows the horse to spread its legs and balance when cornering.

If things go wrong

If you get a flat tyre with horses in the trailer it is usually possible to change a wheel without unloading the horses. Never use a standard jack on a horse trailer that has horses in it. Instead, carry a block of wood that is sloped at each end for the good wheel to be driven onto. This provides a stable platform for the good wheel and lifts the punctured wheel up, enabling it to be changed. Commercial gadgets work just as well but a strong, stable block of wood works fine.

If you hear banging and crashing from the trailer, stay calm and don't slam on the brakes. Ignore the instinct to pull up on a busy road and drop the ramp. It's far too dangerous, as the horse may get loose and cause an accident. Drive until you can find somewhere off the busy road, preferably an enclosed area. Then carefully look in over the ramp or through the windows. **Assess the situation before opening any doors or the ramp.**

Every situation is different, there may be one or more horses in the transporter and you may or may not have a helper. It is extremely dangerous to be inside the trailer or truck with the horses in this situation. If a horse has fallen it may lie still for the moment but will probably start to panic again when it gets its breath back or if it is suddenly released. It is important to try to release the horse or horses without going in. This is where a pocket knife is invaluable. Keep talking to the horse in a calm manner. If a horse is lying down, assess whether it can get up safely by itself. If there is another horse in with the prone one, try to get the standing one out (without going in) so that the prone horse has room to get up. In most situations the prone horse will be able to get up once it has room. If it can't, call a vet.

The horse may have turned around rather than fallen over. If you have a horse trailer with a cover or doors over the ramp the horse can't go anywhere. If the horse can't be turned around again it may be better to carry on with the journey rather than unload the horse in this situation. If the ramp does not have a cover or doors the horse may try to jump over the ramp. It will need to be unloaded and loaded again. Stand to one side of the ramp because the horse will try to come out as the ramp is being lowered. The horse will then need reloading, and you must make sure that it can't get loose again. In situations such as these, assess the state of affairs and make a decision taking into account the safety of humans first and horses second.

Transporting breeding stock

If you are transporting foals the space above the ramp must be fully enclosed. Foals can quickly become frightened, panic and jump out over the ramp. If possible, travel the mare and foal without a centre partition and allow the foal to travel backwards and drink whenever it needs to. If this is not possible the foal will need to feed at least every three hours, maybe more frequently if it is very young.

When loading a mare and foal it is usually easier to put the foal on first and the mare will follow even if she is not usually a good loader. Her maternal instincts will override her principles! Conversely, even a mare that is normally a good loader may not load unless her foal is in front of or beside her as she walks onto the truck or trailer. Again, her maternal instincts dictate her behaviour.

Stallions usually travel well with other horses as long as they are accustomed to the other horses. They should not, however, travel with mares in oestrus. Partitions and head dividers must be used. Be aware that other horses, especially geldings, may be intimidated by the stallion, which will increase stress for all concerned.

Recommended further reading

A good website for information about transporting horses is www.cyberhorse.net.au/safetowing. In particular, there is a comprehensive list of the website addresses of state transport departments.

The Rural Research & Development website (www.rirdc.gov.au) has a copy of the Code of Practice for the Land Transport of Horses that you can download.

10

Selecting horses

Buying or selling a horse may be something that you do infrequently, or you may change horses on a regular basis. Commercial operators often need to purchase and sell horses regularly. Individuals and commercial operators may even lease horses. Whatever the situation, in each case it is important to select the right horse. A leased horse can be returned if it is not suitable. However, if an unsuitable horse is purchased then not only do you have to make time to sell it but if the horse actually has problems that make it unsafe you should not sell it to an unsuspecting buyer (even though this may have been done to you). Sometimes it is possible to return a horse. In the past, when buying a horse it was usually a case of 'buyer beware'. This means that if a person bought a horse and it was not suitable for any reason, there was no going back to the seller. This unwritten rule is beginning to change in our increasingly litigious society. As our culture evolves, it is necessary to consider the use of written contracts when engaging in buying, selling and leasing horses.

When selling a horse, you must ensure that a false impression is not given of the horse. If someone is injured while trying out the horse (under the false impressions) you as the seller may be liable. Likewise, when buying a horse a rider must give a fair impression of their riding ability.

When leasing a horse a written contract should be used to ensure both parties are covered. The contract should be detailed and clear. Standard contracts for all these situations (buying, selling, leasing) can be purchased (see p. 162).

This chapter deals mainly with selecting horses that are to be used for inexperienced riders (either individual owners or commercial operators) because this is where everyone has to start. It could be argued that this is the hardest market to cater for and selection of the wrong horses will become apparent much more quickly. Much of this information is relevant to buyers of other classes of horses and even some very experienced riders have little practice in buying and selling horses.

The process of buying a horse should not be rushed. It usually takes time to find the horse that best suits your needs. An inexperienced person will find it very difficult to buy a good safe horse because it takes experience to tell the difference between a suitable horse and an unsuitable horse. It is also possible to miss a really good horse through indecision, however, it is better to do that than buy the wrong horse. Inexperienced horse people should consider asking an experienced horseperson (such as an instructor) to assist with the purchase. This may mean paying for their time and expenses (if they are professional). It will be money well spent if they are reputable and trustworthy.

Considerations

Budget

Good horses are very valuable. A good beginner's horse is worth its weight in gold to an individual owner or a commercial operation. Inexperienced horse people are usually surprised at how much a good beginner's horse costs. If your budget is limited, you may have to compromise. For example, temperament and training are not areas that a beginner can compromise on but looks, breeding, age and colour are.

Temperament and training

Temperament and training are the most important factors when buying a horse, particularly when it is for an inexperienced rider. On the ground the horse must be friendly, calm and relatively easy to handle. Handlers must be able to feel safe with the horse. The horse should be obedient and not push people around. This type of horse will give a beginner confidence.

It is important that the horse is well-trained in the basics because a well-trained horse is well-behaved. Some horses appear to be very quiet, however, if they do not have good basic training they will not necessarily behave well in a stressful situation. For example, some very young (and therefore relatively inexperienced) horses appear to be very quiet especially if they are in work. Sometimes such a horse is advertised for sale as 'bomb-proof', but in fact the horse is only quiet because it is still growing and is using its available energy for this. Once mature (around five or six years for most horses) a horse may have excess energy and unless it is well-trained it can use this excess energy to misbehave.

Other seemingly quiet horses are so because they are being handled and ridden by experienced people who give clear signals as to what they expect the horse to do. Less experienced people tend to give less clear signals until they have perfected their technique. For this reason they need a horse that is not overly reactive.

A horse that is well-trained in the basics will calmly go forward, stop and move over when asked, both on the ground and under saddle. It will have been habituated to various common situations such as traffic, children, dogs etc. It will behave quietly in company and on its own. The latter is less important for a trail riding operation because the horse would not usually be asked to work alone.

Age

The prime age of a horse is between six and ten years. At this age a sound horse will generally command the best price. For this reason they are often too expensive for many buyers. Older horses that are still sound are a good compromise for inexperienced people as they can still offer many years of service, for example many well-cared-for horses live to be 25-plus. Older horses also tend to be quieter because of their exposure to many sights, sounds and other experiences. An older horse is usually less supple than a young horse and this must be taken into consideration if the horse is going to be checked by a vet.

Young horses and inexperienced riders do not usually mix well. It is not recommended for inexperienced people to train a young horse without intensive experienced support. Some commercial operations prefer to breed and train their own horses so that they can train the horses correctly from the beginning rather than have to buy horses that may already have behavioural problems. This can be a very successful strategy as the horses are being handled and trained by experienced horse people before being used in the business.

Conformation and looks

Conformation and looks are not as important as temperament and training in a beginner's horse. Conformation is the word used to describe how a horse is put together. Whether it is good-looking or not often depends on your perspective. To a lesser or greater extent, the conformation of a horse determines what the horse will be able to do and for how long.

A well-conformed horse that is in proportion and has good legs

Certain conformation faults can be overlooked if a horse has a good temperament and training. Indeed, if the horse has reached a certain age and its poor conformation has not hindered it then probably it is nothing to worry about. Conformation faults in a younger horse should be regarded with more caution, as it is unknown whether they will affect the horse in the long term. Being able to assess a horse's conformation takes experience and a trained eye. An inexperienced horse person will need the advice of an experienced horse person. There are also many good books on the subject if you want to learn how to judge conformation (see p. 162).

In terms of buying a safe beginner's horse, looks are immaterial and indeed many good beginner horses are plain or even downright ugly. These gems are to be prized because of their experience and not their looks

Action and movement

Action and movement are linked to conformation because the way in which a horse is built dictates how it will move. The horse should be able to move freely, without stumbling and tripping, in all the paces. It should move straight without hitting itself (hitting one leg with another). It should be comfortable to ride. In a beginner's horse some imperfections in movement can be forgiven in the same way that imperfections of conformation can be forgiven. As long as the horse is not likely to render itself lame by its imperfect movement then it does not really matter. Again, if an older horse got this far without disabling itself then it should go a lot further.

Good movement is a plus and makes a horse more comfortable to ride, however, flashy big movement is not necessary and may even be a disadvantage in a beginner's horse

Colour

There have always been many myths about colours of horses and associated temperament. There is no proof of any link – it is unwise to link colour to temperament or to disregard or prefer a certain horse just because it is a certain colour. Indeed there is an old saying: 'A good horse is never a bad colour'.

The only factors about colour that may be worth considering are that a lot of white skin in certain areas (such as a blaze that passes over the eye or nostrils) can be susceptible to sunburn and will therefore require more attention in hot weather. White markings in other parts of the body, including white socks which result in white feet, are not usually a problem unless the area in which the horse will be kept is very muddy. White legs tend to be more prone to greasy heel (a skin condition). However, it has been scientifically proven that white feet are *not* weaker than black feet.

Breed

Some breeds are regarded as quieter than others. Selective breeding for good temperament (rather than speed or beauty, for example) has led to some breeds being quieter than others; however, this is not always the case. Not all breeders select breeding stock by temperament. A horse that has a good temperament as well as good looks costs a lot of money, so many breeders can only afford stock that has one trait or the other. Therefore it can be dangerous to believe that a horse will be quiet just because it is a certain breed. There are huge variations of temperament within a breed as well as between breeds.

Thoroughbreds that are 'off the track' are not a good choice for a beginner. Generally, 'hot-blooded' horses such as Thoroughbreds and Arabs are thought to be more excitable than 'cold-blooded' horses such as draught breeds, for example Clydesdales, but there are many exceptions. You should treat every horse as an individual. Each horse has different experiences and different parentage. No horse is automatically well-behaved just because it is a certain breed.

Sex

The sex of the horse is usually immaterial. Some commercial operators stick to horses of one sex (usually geldings) to reduce the chances of fighting (usually geldings fighting over mares). Some people prefer mares and some prefer geldings (see p. 14). It often comes down to that particular person's past experiences. There are good- and bad-tempered mares *and* geldings.

Size

When you are mounted (and with your feet in the stirrups), your foot should ideally be level with the bottom of the horse's stomach or slightly higher. If you are too tall or too short for a particular horse, you will be unable to ride it effectively. Unfortunately this commonly occurs when purchasers fall into the trap of buying a horse that is too large and strong for them.

It is difficult to buy the right size for children as they can grow out of ponies quickly. Try not to be tempted to buy a bigger horse than the child is ready for, as many children

This horse is the correct size for this rider

lose confidence in this situation. On the other hand, many small adults only need a pony or a small horse but actually buy a large horse. They run the risk of overmounting themselves.

Current workload

A horse that is working hard at the time of purchase, such as a riding school or trail riding horse, will not necessarily be equally quiet when it becomes a privately owned horse. In that situation it is usually ridden far less than in a commercial operation. Also, good horses are rarely sold from such establishments because they are simply too valuable.

Conversely, if the horse is not in work at all at the time of purchase it is difficult to ascertain what it will be like when it is working. Even though many sellers will not consider leasing a horse, it is better for a potential buyer to lease for a while to check the horse's behaviour when working. If a seller wishes to command a top price for a horse it should be in regular work when it is put up for sale.

Current condition

The condition of the horse at the time of purchase should be good (not too fat or too thin) if the buyer wants to get a truer picture of the horse's suitability. A horse that is bought in poor condition may behave very differently when it is brought it into good condition. This happens frequently and purchasers are surprised that the timid quiet horse they thought they had bought becomes a different horse when it has been fed properly for a while. Again, a seller should present a horse for sale in good condition.

Methods of buying

There are several methods by which horses are bought and sold. These include horse auctions, dealers, private sales advertised in newspapers, magazines, the internet and word of mouth.

Horse auctions

Buying a horse at an auction has the advantage of enabling buyers to view several horses in one location. However, for inexperienced horse people there are many drawbacks to buying at auction, the main one being that problem horses are often sold at auctions because it enables the owner to dispose of the horse with no comeback.

Spotting the genuine sellers and horses requires skill, and even experts can be caught out. Therefore don't go looking for a beginner's horse at auction if you are inexperienced.

Horse dealers

A good horse dealer is a valuable contact, however, dealers range from those that give a very good service to those that do not. A good dealer provides an excellent way of buying a horse, as they will match you with the right horse and save you time. A horse may be more expensive when purchased from a dealer because that is how they make a living and you are paying for their service. Reputable dealers are aware that if they provide good service you will use their services again in the future and you are likely to recommend them to other people. Ask around before you choose a horse dealer, as the reputation of such individuals is usually common knowledge among horse people.

Private sale

There are lots of horses advertised for sale in newspapers, horse magazines and on the internet and by word of mouth. Many very good horses are sold by word of mouth, their owners preferring their good horse to go to a friend of a friend, for example, rather than a stranger which is what happens if the horse is advertised.

It is a good idea to make inquiries through local people involved in horse activities, such as pony clubs, riding clubs, riding schools, produce stores and saddlery shops.

Buying process

If you are buying, it is a good idea to find out as much as possible about a horse before you travel to see it. This can be done over the phone or via email. If telephoning or emailing in response to an advertisement, be prepared to ask the right questions. Asking the right questions can save hours of unnecessary driving time and petrol wastage. There are several important things that a buyer needs to find out.

- How old is the horse?
- What has the horse done in the past and how much training has the horse had?
- What sort of temperament does the horse have?
- How does the horse behave in terms of transporting, shoeing, catching, riding in traffic, riding around other horses, tying up and clipping etc.?

- Does it have any major conformation faults or past injuries/health problems that may impede its performance?
- Is the horse currently being ridden regularly? If not, how long has it been since it was last ridden? What is the skill level and age of the present rider?
- What is the horse's current condition? Is it underweight or overweight?
- Does the horse have any bad behaviours such as kicking out at other horses, biting or nipping?
- How big is the horse? Has it been officially measured?

There are other questions that you could ask in order to find out more about the horse.

- Why is the horse for sale?
- Is it a gelding or mare?
- How long has the seller owned the horse?
- Where did they get the horse?
- Is it registered and if so, with what society?

When you are the buyer, you need to describe your current riding ability and ask the owner whether they think that the horse would be suitable. Some sellers are very honest and will tell you straight away if they think that the horse would be unsuitable. If so, you should take their advice. But some sellers are less helpful. Even if they say the horse would be perfect for you, that answer (along with many of the answers to the above questions) is only the seller's point of view. It is up to you to thoroughly check the horse before you buy it. The answers to these and any other questions simply give you an idea about the horse and may save you time and petrol if the horse sounds unsuitable at this early stage. At the same time, if the horse does not sound like what you are looking for then say so. Sellers much prefer buyers to be honest rather than waste their time.

Making arrangements to view

When making arrangements to view a horse, make sure that there will be someone who will ride the horse. This is very important. The owner of the horse or one of their party should **always** ride the horse first. Never get on a horse unless one of the seller's party has ridden it first. Even then, only get on if you think you will be able to control the horse.

Viewing a horse

If you are inexperienced, find an experienced horse person to go with you. Avoid taking young children when viewing a horse. Arrive on time and dress appropriately (boots, helmet etc.).

If possible, view the horse being caught. Have the horse walked and trotted on a hard level surface so that you can view the animal's natural action and gait. Along with your experienced helper, look in the horse's mouth to check the age, do a thorough conformation check, feel the legs, look at the feet (ask the owner to pick the feet up), ask about the farrier, notice any lumps and bumps and ask about them. Remember, many older horses will have lumps and bumps. Often these are not serious and do not affect its soundness.

All the time, take mental notes on the temperament of the horse. Also notice the handler's/owner's abilities and attitude. A genuine seller will care where the horse is going and will want a good match. They will also be able to tell you lots about the horse.

Ask to see the registration papers if the horse is registered (not all are or need to be). It is worth mentioning on the phone previously that you will want to see the papers, so that the owner has time to get them ready. The papers will list previous owners, the horse's age and its parentage. The seller should be listed as the owner of the horse; if not, contact the listed owner to check (prior to buying the horse).

If the horse passes these inspections, ask to see it tacked up and ridden by the owner or one of their party. Notice what gear the horse is wearing. If the seller is marketing the horse as suitable for a beginner it should not be wearing a martingale or a severe bit. They should put the horse through its paces – walking, trotting and cantering. If it is a jumper, ask to see this too. Ask them to ride the horse away from the property and notice how it behaves both going away and coming back. Once the seller has demonstrated what the horse is capable of it is time for you to try the horse, if you feel comfortable about doing so. Never be persuaded or cajoled into riding a horse that you are not sure about. Many people have been injured when trying out a horse in this situation, often because they did not like to refuse to ride the horse. Also if your helper advises at this stage that the horse is not suitable there is no point going any further.

Another dilemma when trying out a horse is that the seller's gear and facilities may be unsafe. Do not ride using unsafe gear in unsafe surroundings. If the horse seems suitable, arrange to see it again and take some safe gear with you. Arrange for the seller to take the horse somewhere with better facilities, maybe a local pony club or riding club ground.

Some buyers prefer their experienced helper to ride the horse first. This is fine as long as everyone is willing. Either or both of you need to put the horse through its paces. If it can be arranged with the owner, take the horse on a trail ride away from its home property in the company of one or two other horses.

If the horse has been advertised as good to transport ask to see it loaded and unloaded without fuss into a horse trailer or truck.

If the horse seems suitable after thoroughly checking it out, arrange to see it again soon (if it is near enough) or start negotiating.

Negotiating

Negotiation includes the price (which may or may not be negotiable), whether you can have a trial, the vet inspection and whether there is any sort of guarantee. Some of these issues are best sorted out on the phone before even going to see the horse. For example, ask the owner if they are willing for the horse to be vet checked at your expense (be wary of the horse if they are not). You can ask if they would consider a trial period. Many people will not, quite understandably, but some will.

Often the seller will not allow a trial period, and this is quite reasonable as the question of who is legally responsible for the horse during this period can be a difficult one. If you do enter into a trial agreement use a written contract to protect both you and the seller (see p. 162). If a trial is not permitted, try to return at least one more time to ride and handle the horse before buying. It may be possible to trial the horse at the owner's property for a defined period if the distance is not too great.

Vet checking a horse

Consider having the horse vet checked before buying. Not all people have a horse vet checked and indeed in a commercial situation it is not usually economically viable unless the horse is very expensive.

A veterinary inspection will identify health issues that you should consider before buying a horse. These include sight, teeth, heart function, lung function and skin disease, legs and feet. The vet should carry out a flexion test on each leg. A flexion test involves the leg being held in a flexed position for about one minute and the horse then immediately trotted out by a handler. The vet will watch for lameness in the horse's first steps after release.

Bear in mind that a vet check will not show up everything, it is simply an indication of the horse's general health on that day. Some vets (who specialise in horses) are far more qualified than others. A small animal vet is unlikely to be able to age the horse by its teeth, for example. This is important if the horse is not branded (branding indicates age), as you may be paying an inflated price for a horse that is much older than the seller says it is. Not all sellers deliberately lie about a horse's age. Often they were misinformed when they bought the horse and they do not have the skills to notice that the horse is actually a lot older than they were led to believe.

The vet will need to know what you intend to do with the horse. Vets don't usually declare a horse 'sound', but instead give their opinion on its suitability for the intended purpose.

Some vets will not do vet checks as it can lay them open to liability claims. You may have to ask around to find a vet who will do it. The vet should not be the seller's vet as that puts the vet in a difficult position. A vet can do blood tests to see if the horse is medicated to make it quieter or sounder or both, though this is not common practice except when a large amount of money is changing hands.

Collecting a horse

Arrange transport. Remember that over long distances commercial transporters can work out cheaper than going for the horse yourself. If you do go to pick up the horse it is usually better to let the owner load the horse. The horse knows the owner but not you at this stage.

Consider a written contract on buying the horse. This should include the buyer's and seller's details (names, addresses and telephone numbers), a receipt for money paid, a detailed description of the horse, the date of transaction and a declaration of its attributes, for example if the seller declares the horse to be a good beginner's horse, good to transport etc. they should not object to putting it in writing. Keep a copy of any written advertisement that was used to sell the horse.

Settling a new horse in

Once the new horse is home, introduce it carefully to any other horses. There are many ways that a new horse can be introduced to a group of horses. Remember that the existing group will have a pecking order which the introduction of a new member will disrupt. It is not safe for anyone concerned to simply turn the new member out into a group of horses. The new horse can be run into a fence by the other horses. A safer method is to turn the horse out in an empty paddock then add the other horses one at a time over several hours.

The new horse (spotted horse) is being introduced to an existing herd, one member at a time

Riding a new horse

Once you get the new horse home allow it to settle in before you first ride it. When a horse is moved to a new home it takes time to settle down and behave normally, so don't be surprised if it is not initially as quiet as when you tried it out before buying it. If possible, give the horse a couple of days to settle into the new surroundings before riding it. An older 'been there, done that' type of horse will settle into a new home much more quickly than a horse that has changed homes infrequently or never before. In the latter case the horse has had its world turned upside down and will be unsettled for a while after arrival.

Take things slowly in the beginning. Handle the horse on the ground before riding. Start by riding in an enclosed area and make sure the horse's movements can be controlled. Remember that the horse has to get used to new surroundings as well as a new owner whereas the rider only has to get used to the horse. It is a good idea to have some lessons on a new horse even if the rider does not normally have regular lessons. This will help the rider to work through issues that often arise in this 'getting to know you' stage. When the rider and new horse go for their first ride together, someone should accompany them either mounted or on foot. A rider should not set off on a group ride until they know the horse well.

Selling a horse

Preparation will help with the process of selling. Decide which method you are going to use to sell the horse. Read the previous sections to help you decide.

Many of the issues associated with selling are the same as those for buying, except of course that you are the seller rather than the buyer.

- Prepare the horse for sale before putting it up for sale.
- When composing an advertisement make sure it is directed at the right people. It's a waste of everyone's time, yours included, if your ad gives the wrong impression.
- Plan questions that you would like to ask of callers who respond to the advertisement.
- Find out about the riding ability of potential purchasers before letting them come and try the horse.
- When talking to potential purchasers, tell them if you feel they would be unsuitable for the horse!
- Think about whether you will let the horse go on trial.
- Have the horse caught, clean and in a stable, yard or easily accessible paddock.
- Be prepared to ride the horse or have someone who will do it for you.
- Provide safe gear and facilities for potential purchasers to try the horse out.
- Remember, you have a moral and legal obligation to **not** sell a horse to someone who isn't capable of handling and riding it.

Recommended further reading

Bennet D (1988–1991). *Principles of conformation analysis*, vols 1–3. Fleet Street Publishing, Gaithersburg, MD, US.

Smythe RH & Goody PC (1993). *Horse structure and movement*, 3rd edn. JA Allen, London.

Standard contracts information

Legal contracts can be purchased at saddlery stores or ordered via the internet at www.horseforce.com.au.

Appendix 1

Useful contacts

Official bodies

These organisations are the national (Australian) body address. Each has links to state branches where applicable, as well as numerous useful other links.

- Equestrian Federation of Australia (EFA): www.efanational.com
- Australian Pony Club: www.ponyclub-australia.org
- Australian Horse Industry Council: www.horsecouncil.org.au
- Association for Horsemanship Safety and Education: www.ahse.info
- Australian Horse Riding Centres: www.horseriding.org.au
- Australian Racing Board: www.australian-racing.net.au
- Riding for the Disabled Australia: www.rda.org.au
- Australian Stock Horse Society: www.ashs.com.au
- Harness Racing Australia: www.harness.org.au

International contacts

- American Association for Horsemanship Safety: (AAHS) www.horsemanshipsafety.com
- Equine law and horsemanship safety site (maintained in cooperation with AAHS): www.tarlton.law.utexas.edu/dawson
- American Medical Equestrian Association: www.ameaonline.org
- International Equestrian Federation: www.horsesport.org/fei

Appendix 2
Safe instructing

Good instructors come in all shapes and sizes and from many different backgrounds. Some have formal qualifications and some don't. It is important that all instructors follow safety procedures whenever teaching to ensure the safety of their students and to protect themselves from potential litigation. Safe procedures include those outlined in this book, and those of relevant codes of practice such as the Code of Practice for the Horse Industry (HorseSafe) (July 2003) which is administered by the Australian Horse Industry Council. This code of practice can be downloaded from their website: www.horsecouncil.org.au.

If you would like to become a qualified instructor, start by contacting one of the following organisations.

- The Association for Horsemanship Safety and Education runs clinics and assesses people who are already working in the industry as instructors and trail guides. Successful clients are offered a qualification based on nationally recognised training package units: see www.ahse.info.
- Training in riding instruction is part of the Australian Sports Commission's National Coaching Accreditation Scheme, and the Equestrian Federation of Australia (www. efanational.com) offers three levels of accreditation through this scheme.
- Australian Horse Riding Centres (AHRC) offer two qualifications through the Australian Sports Commission's National Coaching Accreditation Scheme: AHRC NCAS Level 1 (Coaching) and AHRC NCAS Level 1 (Trail Ride). The website www.horseriding.org.au has more information.
- It is possible to become an RDA coach to teach riding at an RDA venue. Contact RDA at www.rda.org.au
- The pony club runs a coach accreditation scheme for their instructors – contact www.ponyclub-australia.org

Recommended further reading
Dawson J (2003). *Teaching safe horsemanship*, 2nd edn. Storey Communications, US.

Glossary

aged horse: an older horse. In the racing industries a horse is regarded as aged by seven to nine years old. In other areas of the industry it usually means at least a teenaged horse.

agistment: providing feed, shelter and water for horses for payment.

all-purpose saddle: a saddle that is suitable for jumping and flat work riding.

arena: a flat area that is designed for working a horse. Can be fenced or unfenced.

bar: part of the upper gum which bears no teeth and takes the bit of the bridle. Also refers to part of the hoof running from the heel alongside frog.

bareback: riding without a saddle.

base narrow: conformation fault describing close-set fore and hind legs.

base wide: conformation fault where the limbs are set wide apart on the body and closer together at the hooves.

baulk: when a horse refuses to go forward.

bell boot: protective cover for the heels and coronary band used during training or competition.

bit: a piece of metal attached to the bridle which runs through the mouth and over the tongue. It is used to aid control of the horse when handling or riding.

blaze: broad white marking covering most of the front of the face.

bot: botfly larva which lives in the horse's stomach. Also refers to the free-living fly stage.

breastplate: a device attached to the saddle that prevents it from slipping back on the horse.

bridle: the part of a horse's saddlery or harness that is placed about the head.

browband: part of a bridle running across the forehead.

buck: movement of a horse's back and legs when it tries to get rid of a rider.

canter: a pace of three-time which is faster than a trot and slower than a gallop.

castration: removal of the testicles of a male horse. The horse is then referred to as a gelding.

cattle grid: a series of metal bars with gaps between designed to discourage cattle from crossing through an open gateway.

cavesson noseband: the standard type of plain noseband.

chaff: hay or oat straw cut into short lengths for use as a feedstuff.

cheekstrap/cheekpiece: adjustable past of the bridle running from headpiece to bit.

clean legs: legs that are free of swellings and deformities.

clench: the part of a nail which, during shoeing, is left projecting from the wall of the hoof after the end of the nail has been twisted off.

clip: removal of hair in horses that are worked through the colder months when the natural hair coat is longer.

cold back: a horse with a stiff back and gait when it is first mounted.

colic: pain in the abdomen.

colt: young male up to four years old.

concentrates: grains etc. that make up the non-roughage part of the feed.

conformation: anatomical arrangement and proportions of parts of the body.

crest: the upper line of a horse's neck.

croup: the upper line of a horse's hindquarters from the highest point to the top of the tail.

cryptorchid: a stallion with one or both testicles retained in the abdomen, also called a rig.

curb chain: a chain fitted to the eyes of a curb or pelham bit and running under the jaw.

curry comb: a piece of grooming equipment used to remove dirt and scurf from a bodybrush. It has a flat back and a front consisting of several rows of small metal teeth.

dock: the bone that runs down the horse's tail.

drench: another term for worming (treatment for parasites).

dressage: the art of training horses to perform all movements in a balanced, supple, obedient and keen manner.

energy: the horse gets energy from feed.

entire: an uncastrated male horse (stallion).

equine: horses are part of the family *equidae*, which includes asses and zebras.

ergot: small horny area in a tuft of hairs behind the fetlock joint.

event horse: a horse which competes or is capable of competing in a combined training competition.

event: a term that covers organised horse shows and other types of event such as an endurance event.

farrier: a person who makes horseshoes and shoes horses.

fibre: the fibrous substance of plants.

filly: young female up to four years old.

flank : the area above and just forward of the back leg of a horse (under the loins).

flexion: movement which bends a part of the body such as a joint.

flight response: a behavioural term used to describe how animals react to danger by fleeing rather than defending themselves.

foal: young horse up to the age of 12 months.

forage: to forage is to graze. The term forage is also applied to fodder that is fibrous.

forehand: the part of the horse which is in front of the rider.

frog: the part of the hoof that, as it comes into contact with the ground, acts as a buffer to absorb the impact and prevent slipping.

full mouth: the mouth of a horse at five years old, when it has grown all its permanent teeth.

gait: sequence of leg movements, usually forward.

gall: a skin sore resulting from rubbing by saddle or girth.

gallop: the fastest gait of a horse.

galloway: a horse 14 to 15 hands high.

gaskin: the part of the hind leg between the stifle and hock.

gelding: a castrated male horse.

girth: the strap that holds the saddle on by running under the horse's chest.

good doer: a horse that gains weight or stays fat on minimal feed.

greasy heel: a skin infection, usually on coronet, heels and pasterns. Also referred to as mud fever.

green horse: a horse that has had basic training and requires further training.

hack: any horse used for riding, also to ride casually rather than in a particular fashion.

hackamore: a bitless bridle.

halter: a rope, nylon or leather headpiece used for leading a horse.

hand: a linear measurement equalling 4 inches (approximately 10 cm), used for giving the height of a horse from the ground to its withers.

hard feed: concentrates fed to horses, i.e. grain or mixes.

headpiece: part of the bridle which goes over the top of the horse's head and behind the ears.

hobbles: leather, rope or chain used to restrain a horse by restricting leg movement.

hock: joint in the hind leg between the gaskin (second thigh) and cannon bone.

hood: a head cover which has openings for ears and eyes.

hoof pick: a hooked metal instrument used for removing stones and dirt from a horse's foot.

horse person: a person who has regular contact with horses.

horse: a general tem that includes ponies (see **pony**). Sometimes used to describe a stallion (as opposed to a mare).

horsemanship: a traditional term used to describe a person's general skills with horses, i.e. riding and handling skills.

irons: stirrup irons.

jog: a slow short-paced trot.

lameness: unevenness of the horse's stride when moving; limping.

left rein: to be 'on the left rein' means moving to the left.

loins: lower back area behind the saddle.

long reining: a stage of breaking involving driving a horse from the ground.

lunge: to exercise a horse on the end of a rope in a circle.

mare: a female horse more than four years old.

martingale: a device used to stop a horse from holding its head too high when ridden.

mouthing: educating a horse to respond to pressure on the reins. Also refers to determining the approximate age of a horse by looking at its teeth.

muck out: to clean out a stable of droppings and dirty bedding.

mud fever: a skin infection of heels and pastern. See **greasy heel**.

near side: the left-hand side of a horse.

nipping: a horse giving a small bite.

novice : for the purpose of this book, a novice is someone with limited horsemanship experience.

off side: the right-hand side of a horse.

parasite: an organism that lives on or inside the horse.

pastern: the portion of the leg between the fetlock and the hoof.

poll: highest point of the head, just behind the ears.

pony: a horse of any breed up to 14 hands.

quarters: the area of a horse's body extending from the rear of the flank to the root of the tail and downwards to the top of the leg; the hindquarters.

quidding: collecting feed in the mouth without swallowing it, often associated with teeth problems.

racing saddle: a saddle that is designed specifically for racing and training race horses.

rear: a horse rising up on its hind legs.

rein back: to make a horse step backwards while being ridden or driven.

reins: pair of long narrow straps attached to the bit or bridle, used by the rider or driver to guide and control the horse.

riding instructor: someone who instructs other people to ride (may be qualified or unqualified).

rig: abnormally developed or improperly castrated male. See also **cryptorchid**.

roller: a special girth with rings at the top through which long reins can pass.

roughage: high-fibre feed such as pasture, hay and chaff.

running martingale: y-shaped strap which connects the girth to each of the reins. The reins run through metal rings on the end of the strap.

saddle cloth: padding under the saddle.

saddle tree: the foundation on which a saddle is built.

saddler: a person who makes or deals in saddlery and/or harness.

schoolmaster/schoolmistress: a horse that is so experienced and educated that it will enable a beginner rider to learn to ride safely.

shy: jump or move sideways or backwards at an unexpected sight or noise that frightens the horse.

snaffle bit: the oldest and simplest form of bit, consisting of a single bar with a ring at each end to which one pair of reins is attached.

soundness: state of health or fitness to carry out a particular function.

spur: a pointed device strapped onto the heel of a rider's boot and used to urge the horse onwards.

stable: a building that horses live in.

stallion: an ungelded male horse aged four years or over.

stirrup leathers: straps that connect the stirrups to the saddle.

stirrup rubbers/treads: inserts into a stirrup iron that provide extra grip.

strangles: an infectious and highly contagious disease caused by the bacteria *Streptoccus equi.*

studs: screw-in projections in the heel of a shoe to give traction. Stud also refers to a place where horses are kept for breeding.

supplement: extra feed other than pasture.

surcingle: a belt of webbing or leather that passes over a saddle and girth. It is used to keep the saddle in position.

tack: saddlery/gear.

tetanus: an often-fatal disease related to wounds.

throat latch: a strap that is part of the headpiece of a bridle.

tinker's grip: a restraining hold. You grip a fold of skin on the horse's neck to settle a fractious horse.

trackrider: a person that exercises racehorses for a living.

trail ride: to go out for a ride rather than ride in an arena.

trot: a pace of two-time in which the horse's legs move in diagonal pairs simultaneously.

twitch: a restraining hold. Pressure is placed on a horse' nose with a twisted rope or a hand.

unsound: a horse with any defect that makes it unable to function properly.

vaccinate: to inject vaccine to stimulate immunity, e.g. for tetanus.

vice: a common term for horses that display stereotypic behaviours, e.g. windsucking.

walk: the slowest of the paces: it has four beats.

weanling: a young horse which has been weaned from its mother.

withers: top of the shoulders between the neck and back.

wolf tooth: a small vestigial pre-molar tooth which is often removed.

yearling: horse between one and two years old.

Bibliography

Australian Horse Industry Council (2003). *Code of practice for the horse industry.* (Horse*Safe*). Australian Horse Industry Council.

Brady S (2005). *Horse training: Steve Brady's formula for success.* Steve and Linda Brady, NSW, Australia

Cargill C (1999). *Reducing dust in horse stables and transporters.* RIRDC, Canberra.

Chatterton J (2000). *John Chatterton's ten commandments.* John & Janet Chatterton, Queensland, Australia.

Cripps R & Pagano H (2005). *Monitoring falls during eventing.* Report for the Rural Industries Research & Development Corporation, RIRDC, Canberra.

Dawson J (1997). *Teaching safe horsemanship.* Storey Communications, US.

EFA website: www.efanational.com/content/risk_management/safety

Gunning L (2003). After a fall. *Hoofbeats Magazine*, August/September.

Huntington P, Myers J & Owens L (2004). *Horse sense: the guide to horse care in Australia and New Zealand*, 2nd edn. Landlinks Press, Melbourne.

Managing health and safety in the racing industry (Queensland).

McLean A & McLean M (2002). *Horse training the McLean way: the science behind the art.* Australian Equine Behaviour Centre, Victoria, Australia.

Sport & Recreation Victoria. *Preventing equestrian injuries: facts on horse-related injuries.* Dept of Sport & Recreation, Melbourne. See www.monash.edu.au/muarc/projects/sport.html.

Vicroads website (road safety, horses and traffic): www.vicroads.vic.gov.au.

Watts GM & Finch CF (1996). Locking the stable door: preventing equestrian injuries. *Sports Medicine* 22 (3): 187–197.

Watts SM (2003). Falling off. *Hoofbeats Magazine*, August/September.

Index

HORSE SAFE

A COMPLETE GUIDE ~~WITHDRAWN~~ E SAFETY

JANE MYERS

LAND

National Library of Australia Cataloguing-in-Publication entry:

Myers, Jane.
  Horse safe.

  Bibliography.
  Includes index.
  ISBN 0 643 09245 5.

  1. Horses - Safety measures. 2. Horsemanship - Safety measures. I. Title.

  798.230289

Published by and available from:
Landlinks Press
150 Oxford Street (PO Box 1139)
Collingwood Vic 3066
Australia

Telephone: +61 3 9662 7666
Local call: 1300 788 000 (Australia only)
Fax: +61 3 9662 7555
Email: publishing.sales@csiro.au
Website: www.landlinks.com

Landlinks Press is an imprint of **CSIRO** PUBLISHING

Cover
Front cover photo by Stuart Myers; back cover photos by the author

Set in 10.5/13 Minion
Cover and text design by James Kelly
Edited by Adrienne de Kretser
Index by Russell Brooks
Typeset by J&M Typesetting
Printed in Australia by Ligare